P9-CAG-747

Monograph Series

MARKERS IN CARDIOLOGY:

Current and Future Clinical Applications

Previously published in the American Heart Association Monograph Series:

Cardiovascular Applications of Magnetic Resonance
Edited by Gerald M. Pohost, MD

Cardiovascular Response to Exercise
Edited by Gerald F. Fletcher, MD

Congestive Heart Failure: Current Clinical Issues
Edited by Gemma T. Kennedy, RN, PhD and Michael H. Crawford, MD

Atrial Arrhythmias: State of the Art
Edited by John P. DiMarco, MD, PhD and Eric N. Prystowsky, MD

Invasive Cardiology: Current Diagnostic and Therapeutic Issues
Edited by George W. Vetrovec, MD and Blase Carabello, MD

Syndromes of Atherosclerosis: Correlations of Clinical Imaging and Pathology
Edited by Valentin Fuster, MD, PhD

Pathophysiology of Tachycardia-Induced Heart Failure
Edited by Francis G. Spinale, MS, MD, PhD

Exercise and Heart Failure
Edited by Gary J. Balady, MD and Ileana L. Piña, MD

Sudden Cardiac Death: Past, Present, and Future
Edited by Sandra B. Dunbar, RN, DSN, Kenneth A. Ellenbogen, MD, and Andrew E. Epstein, MD

Hormonal, Metabolic, and Cellular Influences on Cardiovascular Disease in Women
Edited by Trudy M. Forte, PhD

Discontinuous Conduction in the Heart
Edited by Peter M. Spooner, PhD, Ronald W. Joyner, MD, PhD, and José Jalife, MD

Frontiers in Cerebrovascular Disease: Mechanisms, Diagnosis, and Treatment
Edited by James T. Robertson, MD and Thaddeus S. Nowak, Jr., PhD

Current and Future Applications of Magnetic Resonance in Cardiovascular Disease
Edited by Charles B. Higgins, MD, Joanne S. Ingwall, PhD, and Gerald M. Pohost, MD

Pulmonary Edema
Edited by E. Kenneth Weir, MD and John T. Reeves, MD

The Vulnerable Atherosclerotic Plaque: Understanding, Identification, and Modification
Edited by Valentin Fuster, MD, PhD

Obesity: Impact on Cardiovascular Disease
Edited by Gerald F. Fletcher, MD, Scott M. Grundy, MD, PhD, and Laura L. Hayman, PhD, RN, FAAN

The Fetal and Neonatal Pulmonary Circulations
Edited by E. Kenneth Weir, MD, Stephen L. Archer, MD, and John T. Reeves, MD

Ventricular Fibrillation: A Pediatric Problem
Edited by Linda Quan, MD and Wayne H. Franklin, MD

Monograph Series

MARKERS IN CARDIOLOGY:

Current and Future Clinical Applications

Edited by

Jesse E. Adams III, MD
Investigator
Jewish Hospital Heart and Lung Institute
Assistant Clinical Professor of Medicine
Division of Cardiology
University of Louisville
Louisville, Kentucky

Co-edited by

Fred S. Apple, PhD
Medical Director
Clinical Laboratories
Hennepin County Medical Center
Professor of Laboratory Medicine and Pathology
University of Minnesota School of Medicine
Minneapolis, Minnesota

Allan S. Jaffe, MD
Senior Assistant Consultant
Mayo Clinic
Professor of Medicine
Cardiovascular Division
Mayo Medical School
Rochester, Minnesota

Alan H.B. Wu, PhD
Professor
Department of Pathology and Laboratory Medicine
Hartford Hospital
Hartford, Connecticut

Futura Publishing Company
Armonk, NY

Library of Congress Cataloging-in-Publication Data

Markers in cardiology : current and future clinical applications / edited by Jesse E. Adams III ; co-edited by Allan S. Jaffe, Fred S. Apple, Alan H.B. Wu.
p. ; cm.
Includes bibliographical references and index.
ISBN 0-87993-472-7
1. Myocardium—Diseases—Diagnosis. 2. Biochemical markers. I. Adams, Jesse E.
[DNLM: 1. Coronary Disease—diagnosis. 2. Biological Markers. 3. Coronary Disease—physiopathology. 4. Troponin—diagnostic use.
WG 300 M345 2001]
RC685.M9 M366 2001
616.1′24075—dc21

00-052792

Published by
Futura Publishing Company, Inc.
135 Bedford Road
Armonk, New York 10504

LC #: 00-052792
ISBN #: 0-87993-472-7

Printed in the United States of America on acid-free paper.

Preface

Diagnostic accuracy is critical for the optimal diagnosis and treatment of patients with cardiac disease. However, correctly identifying patients who have cardiac disease can be a challenge for even the most astute clinician. Furthermore, the identification of those patients who are likely to have deleterious cardiac events or are less likely to respond to therapies is even more difficult. Accordingly, an ever-increasing reliance has been placed on the measurement of diverse proteins that can indicate the presence of cardiovascular disease and help to identify optimal therapeutic alternatives. Initially used primarily to aid in the diagnosis of cardiac injury, markers are now being used with increasing frequency to triage patients or to guide therapy. However, the optimal application of markers in diagnostic and treatment algorithms requires an understanding of not only the advantages but also the potential limitations.

Our understanding of these protein markers has been advancing rapidly. Academic and clinical cardiologists, laboratorians, and emergency physicians have all contributed to our increased understanding of the proper application of markers to patients with cardiac disease. This monograph contains chapters written by many of the experts in the field, and is specifically designed for physicians or researchers who wish to understand some of the seminal issues in the clinical application of protein markers to patients with cardiac disease.

Acknowledgments

We wish to thank Stan Goldman, PhD, for his editorial review and for his insistent, persistent, and consistent correspondence with the contributing authors, and Joanna Levine of Futura for preparing the manuscript for publication. We also wish to thank the contributing authors, without whom this monograph would not have been possible, and Stuart Horowitz, PhD, Director of Research and Technology at Jewish Hospital, Louisville, KY; Doug Shaw, President of Jewish Hospital; and John Oldfather, PhD, Vice President of Jewish Hospital for their continuing involvement in research on markers in cardiology and for their support of this monograph.

Contributors

Jesse E. Adams III, MD, FACC Investigator, Jewish Hospital Heart and Lung Institute; Assistant Clinical Professor of Medicine, Division of Cardiology, University of Louisville, Louisville, KY

Fred S. Apple, PhD Medical Director, Clinical Laboratories, Hennepin County Medical Center; Professor of Laboratory Medicine and Pathology, University of Minnesota School of Medicine, Minneapolis, MN

Stephanie W. Bateman, BA Research Assistant, Dartmouth Medical School, Lebanon, NH

Linda Cise, MS Research Technician, Roudebush VAMC, Indianapolis, IN

Désiré Collen, MD, PhD Professor of Medicine, Director of the Center for Molecular and Vascular Biology and Department of Cardiology, University of Leuven, Leuven, Belgium

George K. Daniel, MD Cardiology Fellow, Krannert Institute of Cardiology, Indiana University School of Medicine and Roudebush VAMC, Indianapolis, IN

Rose Felten, MS Research Technician, Krannert Institute of Cardiology, Indiana University School of Medicine and Roudebush VAMC, Indianapolis, IN

V.J. Ferrans, MD, PhD Pathology Section, National Heart, Lung, and Blood Institute, National Institutes of Health, Bethesda, MD

George A. Fischer, PhD, DABCC Technical Director, Chemistry, Brigham and Women's Hospital, Boston, MA

Norbert Frey, MD Medizinische Klinik II, Medizinische, Universität zu Lübeck, Lübeck, Germany

Robert Fromm, MD, MPH Associate Professor of Medicine, Department of Medicine, Section of Cardiology, Baylor College of Medicine, Houston, TX

Evangelos Giannitsis, MD Medizinische Klinik II, Medizinische, Universität zu Lübeck, Lübeck, Germany

W. Brian Gibler, MD Richard C. Levy Professor of Emergency Medicine and Chair of the Department of Emergency Medicine, University of Cincinnati College of Medicine; Director of Center for Emergency Care, University of Cincinnati Hospital, Cincinnati, OH

Jessica Gillespie, BS Krannert Institute of Cardiology, Indiana University School of Medicine and Roudebush VAMC, Indianapolis, IN

Jan F.C. Glatz, PhD Associate Professor of Physiology, Cardiovascular Research Institute Maastricht (CARIM); Department of Physiology, Maastricht University,
Maastricht, The Netherlands

Britta U. Goldmann, MD Cardiology Department, University Hospital Eppendorf, Hamburg, Germany

David A. Grundy, MD Emergency Medicine Training Program, Department of Emergency Medicine, University of Cincinnati College of Medicine, Cincinnati, OH

Richa Gupta, MD Postdoctoral Fellow, Krannert Institute of Cardiology, Indiana University School of Medicine and Roudebush VAMC, Indianapolis, IN

Eugene H. Herman, PhD Division of Applied Pharmacology Research, Center for Drug Evaluation and Research, Food and Drug Administration, Laurel, MD

Wim T. Hermens, PhD Cardiovascular Research Institute Maastricht (CARIM); Professor of Molecular Biophysics, Maastricht University, Maastricht, The Netherlands

Paul Holvoet, PhD Centre for Experimental Surgery and Anesthesiology, Katholieke Universiteit Leuven, Leuven, Belgium

Michael P. Hudson, MD Duke Clinical Research Institute, Durham, NC

Allan S. Jaffe, MD Senior Assistant Consultant, Mayo Clinic; Professor of Medicine, Cardiovascular Division, Mayo Medical School, Rochester, MN

Robert L. Jesse, MD, PhD Associate Professor, Internal Medicine/Cardiology, Director, Acute Cardiac Care, Medical College of Virginia Campus, Virginia Commonwealth University, Richmond, VA

Saeed A. Jortani, PhD, FACB Senior Research Associate, Director, Diagnostic Reference Laboratory, Department of Pathology and Laboratory Medicine, University of Louisville School of Medicine, Louisville, KY

Hugo A. Katus, MD Medizinische Klinik II, Medizinische, Universität zu Lübeck, Lübeck, Germany

Michael C. Kontos, MD Assistant Professor, Internal Medicine/Cardiology, Associate Director, Acute Cardiac Care, Medical College of Virginia Campus, Virginia Commonwealth University, Richmond, VA

Ralf Labugger, MSc PhD Candidate, Department of Physiology, Queen's University, Kingston, Ontario, Canada

Johannes Mair, MD Associate Professor of Clinical Chemistry and Laboratory Medicine, University of Innsbruck; Specialist in Clinical Chemistry and Laboratory Medicine, Resident in Internal Medicine and Cardiology, Department of Internal Medicine, Division of Cardiology, University Hospital of Innsbruck, Innsbruck, Austria

Jason L. McDonough, BSc(H) PhD Candidate, Department of Physiology, Queen's University, Kingston, Ontario, Canada

Vickie A. Miracle, RN, EdD, CCRN, CCNS Director of Education, Jewish Hospital Heart and Lung Institute, Louisville, KY

Margit Müller-Bardorff, MD Medizinische Klinik II, Medizinische, Universität zu Lübeck, Lübeck, Germany

E. Magnus Ohman, MD Duke Clinical Research Institute, Durham, NC

Kelly S. Quinn-Hall, MT (ASCP) Technical Specialist, Dartmouth-Hitchcock Medical Center, Lebanon, NH

Paul M. Ridker, MD, MPH, FACC Associate Professor of Medicine, Harvard Medical School, Brigham and Women's Hospital, Boston, MA

Michael E. Ritchie, MD Associate Professor of Medicine, Krannert Institute of Cardiology, Indiana University School of Medicine; Chief, Section of Cardiology, Roudebush VAMC, Indianapolis, IN

Robert Roberts, MD Don W. Chapman Professor of Medicine, Professor of Medicine and Cell Biology, Department of Medicine, Section of Cardiology, Baylor College of Medicine, Houston, TX

Kathy Sturdevant, BS Research Technician, Roudebush VAMC, Indianapolis, IN

Roland Valdes, Jr., PhD, FACB Professor and Vice-Chairman, Director, Clinical Chemistry and Toxicology, Department of Pathology and Laboratory Medicine, University of Louisville School of Medicine, Louisville, KY

Frans Van de Werf, MD, PhD Professor of Medicine, Head of the Department of Cardiology, University of Leuven, Leuven, Belgium

Jennifer E. Van Eyk, PhD Assistant Professor, Department of Physiology, Queen's University, Kingston, Ontario, Canada

Johan Vanhaecke, MD, PhD Professor of Medicine, Head of Heart Failure/Heart Transplantation Unit, Department of Cardiology, University of Leuven, Leuven, Belgium

Britta Weidtmann, MD Medizinische Klinik II, Medizinische, Universität zu Lübeck, Lübeck, Germany

Stacey Wieczorek, PhD Postdoctoral Fellow, Hartford Hospital, Hartford, CT

Alan H.B. Wu, PhD Professor, Department of Pathology and Laboratory Medicine, Hartford Hospital, Hartford, CT

Kiang-Teck J. Yeo, PhD Associate Professor, Dartmouth Medical School and Dartmouth-Hitchcock Medical Center, Lebanon, NH

Contents

Chapter 1

Analytical Issues Affecting the Clinical Performance of Cardiac Troponin Assays

Alan H.B. Wu, PhD

Introduction

Cardiac troponins T (cTnT) and I (cTnI) represent a new generation of biochemical markers that have clinical utility in patients with ischemic heart disease. Research studies have shown that assays for cTnT and cTnI have higher clinical sensitivity and specificity for the detection of ischemic myocardial injury than do the standard serum markers such as creatine kinase (CK) and the CK-MB isoenzyme. Despite these very promising clinical studies, some cardiologists and laboratorians are reluctant to abandon CK-MB because of many unresolved issues. In this chapter we review some of these issues, including the lack of a thorough understanding of how and which specific troponin subunits are released after injury, the lack of assay standardization for cTnI, confusion as to the specificity of different "generations" of troponin assays, differences between assay performance for cTnT and cTnI, the frequency of false positives, confusion about the proper use of decision limits for risk stratification, and poor interassay precision at these limits.

Release of Cardiac Troponin Subunits after Myocardial Injury

Troponin is a complex of three proteins, C, T, and I, that regulate muscle contraction. Troponin T (molecular weight 37,000 d) is so named because it anchors the complex to tropomyosin of the thin fila-

From: Adams JE III, Apple FS, Jaffe AS, Wu AHB (eds). *Markers in Cardiology: Current and Future Clinical Applications.* Armonk, NY: Futura Publishing Company, Inc.; © 2001.

ment, whereas troponin I (24 kd) inhibits actomyosin ATPase, and troponin C (18 kd) is a calcium-binding subunit. There is clinical interest in measuring cTnT and cTnI because these isoforms differ structurally from their skeletal muscle counterparts. Thus cardiac troponin assays should have high specificity for myocardial injury. Although the majority of cTnT and cTnI resides bound to the contractile apparatus, about 6% to 8% of cTnT and 2.8% to 4.1% of cTnI are found in the cytosol.[1] After myocardial damage, the free subunits are putatively the first to be released into blood. This first release is followed by the release of the intact complex from degrading heart myofibrils and its subsequent degradation into smaller complexes, free subunits, and fragments. Protein characterization studies have shown that free troponin T and the intact complex of troponin T-I-C can be detected in serum.[2] In contrast, the major form of troponin I is the binary I-C complex with few free cTnI subunits.[3,4] One postulate is that due to its hydrophobic nature, free cytosolic cTnI released from damaged myocytes binds to existing troponin C circulating in serum to form the binary complex.[5] These subtle differences in release and appearance kinetics may explain some differences in the clinical performance of these markers.

Evolution of Commercial Cardiac Troponin Assays

The first commercial cardiac troponin assay was approved by the US Food and Drug Administration (FDA) in the early 1990s for cTnT (Cardiac T ELISA®) by Boehringer Mannheim Corp (now Roche Diagnostics, Indianapolis, IN), but it suffered from nonspecific binding of skeletal muscle troponin T (sTnT).[6] In the mid 1990s, this assay was replaced with the second-generation cTnT assay (Cardiac T Enzymun-Test®), whereby the cross-reactivity with sTnT was eliminated through replacement of the monoclonal antibodies used in the assay.[7] A third-generation cTnT assay was recently introduced with modifications to the assay calibrators. Since there was no change in antibodies, the specificity of this assay was the same as that of the previous generation test. Due to patent restrictions, there are currently no other manufacturers of approved cTnT assays.

The first two cTnI assays to be developed were the Stratus (Dade) and Opus (Behring, now Dade Behring, Deerfield, IL). The Stratus assay has undergone some modifications to be compatible with the release of new instrumentation (Dimension DSL and Stratus CS), and the Opus assay is being discontinued. Other manufacturers, including Abbott (Abbott Park, IL), Bayer Diagnostics (Tarrytown, NY), Beckman Coulter (Fullerton, CA; formerly Sanofi Pasteur), Diagnostic Products Corporation (Los Angeles, CA), Ortho-Clinical Diagnostics, a Johnson &

Johnson company (Raritan, NJ), and Tosoh Corporation (Tokyo, Japan), have released cTnI assays. Second-generation assays for cTnI have been released or are being developed for many of these platforms.

Lack of Assay Standardization

A major continuing issue with commercial cTnI assays is the lack of assay standardization. In an examination of published data and manufacturer package inserts, slope biases between commercial cTnI assays and the Stratus ranged from a low of 0.10 for the Beckman Access to a high of 3.5 for the Abbott AxSYM, a 35-fold difference.[8] This lack of agreement makes it very difficult if not impossible to apply research data to clinical practice unless the assay used is identical in generation and kind to the published report. Moreover, one cannot use data from one hospital to another on the same patient unless the same assay is used. The American Association for Clinical Chemistry has created a subcommittee charged with preparing a reference material to be used in standardizing results of cTnI assays. Although application of a universal standard will not entirely eliminate the biases since there is no plan to standardize the antibodies used in each manufacturer's kit, differences between cTnI assay results should be greatly reduced to a onefold difference or less between each other.

Specificity Issues for Troponin Assays

Renal Failure

Early clinical studies demonstrated that the first generation cTnT assay produced a high incidence of false-positive results in patients with chronic renal failure[9,10] and skeletal muscle myopathies.[11,12] The second-generation cTnT assay resulted in a significant reduction in the incidence of false-positive results in patients with chronic renal failure.[13] Since the inception of commercial assays, the incidence of abnormally high concentrations of cTnI has been lower than that of cTnT, even when using the second-generation cTnT.[13,14] If high cTnT concentrations reflect cardiac injury, what is the cause for the lower incidence of cTnI in patients with renal failure? Three hypotheses can be considered: 1) there is re-expression of cTnT in regenerating skeletal muscle and renal tissue; 2) cTnT is a more sensitive marker than cTnI for detecting minor myocardial injury; and 3) cTnT epitopes remain in the blood of renal failure patients longer than do cTnI epitopes after true cardiac injury. Independent studies conducted by Ricchiuti and colleagues[15] showed that cTnT isoforms were expressed in regenerating noncardiac tissue, but they did not cross-react with antibodies used in commercial cTnT assays. Haller and colleagues,[16] however, found no expression of cTnT in regenerating tissue. Thus, the prevailing view today is that high cTnT concentrations in renal failure patients are a reflection of true myocardial injury. Support for the second hypothesis comes from the fact that the tissue distribution of cTnT at 10.8 mg/g wet weight is greater than that of cTnI at 4 to 6 mg/g wet weight.[1] If two assays have the same analytical sensitivity for measuring a given quantity of protein, the marker that is released in higher concentrations will be more sensitive for detecting minor myocardial injury. The selection of antibodies used in cardiac troponin assays could also play a major role in detection of residual troponin subunits, to support the third hypothesis. Gel filtration studies have shown that the assay for cTnT detects complexes, free subunits, and fragments that are smaller than the intact cTnT subunit.[2] But because the concentration of free cTnI is low[3,4] and current assays appear to not be sensitive to cTnI fragments,[2] the incidence of finding troponin I in renal failure patients would be expected to be lower. A combination of the last two hypotheses may be responsible for the observation of increased incidence of abnormal cTnT relative to cTnI in renal failure.

The clinical significance of finding cardiac troponin in patients with renal failure remains to be determined. Some studies have demonstrated that increased concentrations of cTnT may be predictive of poor outcomes in renal patients.[17] Other studies have failed to demonstrate risk stratification utility.[18,19] In the study by Van Lente and colleagues,[20] odds ratios for poor outcomes using cTnT and cTnI for patients presenting with suspected acute coronary syndromes and renal insufficiency were lower than for suspected coronary patients without renal insufficiency. In each of these studies, the total number of patients studied was small (<60 each). Larger clinical trials are needed before firm conclusions can be reached about the clinical significance of increased troponin concentrations in patients with renal disease.

False Positives: Fibrin, Rheumatoid Disease, and Human Anti-mouse Antibodies

A source of major concern for troponin assays is the frequency of false-positive results due to the presence in serum of fibrin clots, heterophile antibodies, and human anti-animal antibodies. Fibrin strands and clots form when non-anticoagulated blood is centrifuged, and the serum is removed before there is complete retraction of the clot. Because cardiac patients are routinely given intravenous heparin, in vitro clotting times are prolonged. With the demands placed on the clinical staff to reduce turnaround times for results of laboratory tests (including cardiac markers), it is common for an analyst to place a blood sample into the immunochemistry analyzer while the clotting process is still occurring. The presence of fibrin strands can produce a false-positive result through the incomplete separation of analyte-bound signal antibodies from labeled antibodies free of the analyte. Figure 1 illustrates a possible scenario as to how this interference can occur for one commercial troponin assay. The interference by fibrin can be eliminated or greatly reduced by use of plasma or with thorough centrifugation of the sample before analysis.[21]

The presence of unusual antibodies in a serum sample can also produce both false-positive[22,23] and false-negative[24] results for cardiac troponin. Heterophile antibodies are weak multispecific antibodies that are often present in patients with rheumatoid arthritis. Human anti-animal antibodies are produced when an individual is exposed to ani-

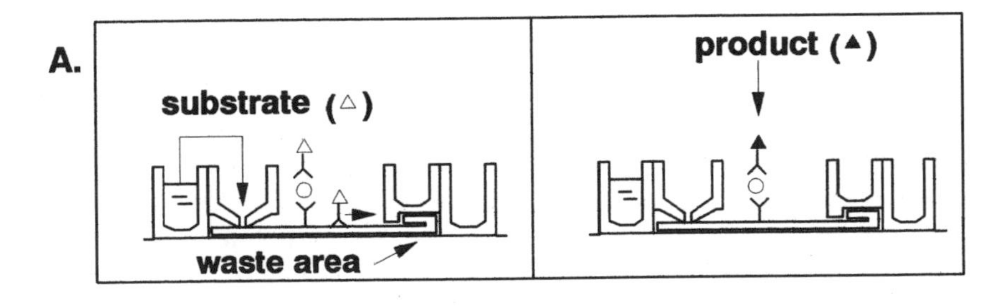

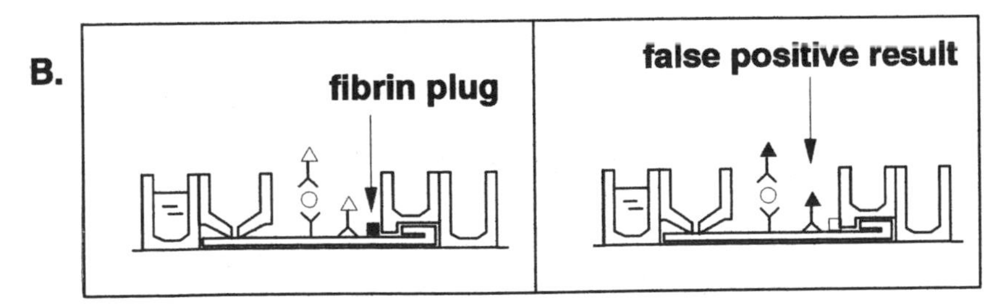

Figure 1. Possible effect of fibrin strands for producing false-positive results on an immunoassay analyzer. A. The analyte (○) is captured by the first antibody (Y). The second enzyme-labeled antibody also binds to the analyte. When the substrate (△) to the enzyme is added, a colored product (▲) is formed. The unbound signal antibody moves to the waste area. B. The presence of a fibrin plug (■) may block the movement of unbound labeled antibodies to the waste area. In this case, the addition of substrate can produce a falsely positive signal.

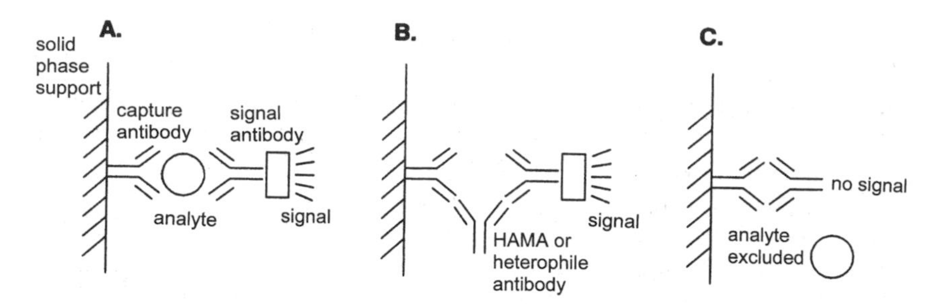

Figure 2. Interference in troponin assays. A. Normal sandwich immunoassay where analyte (○) binds to capture and detection antibodies to produce a signal. B. The presence of heterophile antibodies or human anti-mouse antibodies (HAMA) binds to both the capture and detection antibodies to produce a false-positive signal in the absence of the analyte. C. Heterophile antibodies can also produce a false-negative result by binding to the capture antibody only, thereby excluding the analyte for detection.

mal antigens through a specific occupation (veterinary science) or by animal-specific immunizations (for example, monoclonal antibody drug therapy).[25] Unlike heterophiles, these antibodies bind to antigens from specific species. Human anti-mouse antibodies (HAMA) will only bind to monoclonal antibodies that are raised from murine monoclonal cell lines. The effect of atypical antibodies in blood is illustrated in Figure 2. The normal sandwich immunoassay requires binding of the analyte to both the capture and detection antibodies (Fig. 2A). If either a heterophile or HAMA is present, it will mimic the analyte by bridging between the capture and detection antibodies, thus producing a false-positive signal (Fig. 2B). A false-negative result will be produced if the interfering antibody binds to the capture antibody alone (Fig. 2C). The susceptibility of specific cardiac troponin assays to these interferences depends on the type of antibodies used and the amount of "blocking" reagents, if any, added to the assay formulation. The addition of mouse immunoglobulin G, for example, can be useful in eliminating interferences, as HAMA binds to the blocking reagent instead of the analyte antibodies. Table 1 lists the designs of current commercial troponin assays. All assays are potentially subject to heterophile interferences. Those that use dual monoclonal antibodies are further subjected to interferences by HAMA. The FDA does not require manufacturers to provide specific information as to the identity or type of blocking reagents used.

Table 1

Commercial Assay Formats for Cardiac Troponin T and I Assays

Mono/Mono	Mono/Poly (Goat)	Other Formats
Roche cTnT	Abbott AxSym	Spectral (poly mouse/poly rabbit)
Tosoh AIA	Bayer ACS:180	DPC Immunolite (mono/poly bovine calf)
Stratus/DSL	Bayer Immuno 1	Dade Opus (poly goat/poly goat)
First Medical	Biosite Triage	
Beckman Access		

Assay Cut-off Concentrations for Risk Stratification of Unstable Angina

Since the initial studies for cTnT[26] and cTnI,[27] there have been dozens of studies that have confirmed the notion that highly sensitive and specific biochemical markers such as cardiac troponin can be used to risk stratify patients with unstable angina (that is, a high troponin concentration signifies high short-term risk for future cardiac events). Although the initial studies were mostly on the use of cTnT, the literature on cTnI is increasing. There have now been at least five studies that have compared cTnT with cTnI for risk stratification in the same patient population.[28–32] The cumulative meta-analysis for the five studies, listed in Table 2, shows that cTnI has a slightly higher odds ratio than does cTnT, but the difference is not statistically significant. One study, which showed that cTnT was significantly superior to cTnI, was not included in this analysis because an early blood sample (<12 h) was used in the comparisons.[33] Cardiac TnT may have an advantage for early blood collections because it may be increased sooner after

Table 2

Meta-analysis for the Comparison of cTnT versus cTnI for Risk Stratification

Reference	n	Odds Ratio, cTnT	Odds Ratio, cTnI
Luscher et al[28]	516	3.4 (1.7–6.9)	2.6 (1.3–5.2)
Hamm et al[29]	773	25.8 (9.6–49)	61 (15–512)
Olatidoye et al[30]	107	16.7 (3.4–81)	21.7 (4.3–110)
Green et al[31]	401	3.0 (1.9–4.8)	3.9 (2.4–6.3)
Ottani et al[32]	74	1.6 (0.5–45)	3.2 (1.5–6.5)
Cumulative	1973	5.7 (3.9–8.4)	7.5 (4.9–11.4)

injury than is cTnI. These same investigators showed that collection of later samples improves the risk stratification capabilities of cardiac troponin.[34]

A major issue in these research studies was the determination of the appropriate cut-off concentration for the detection of minor myocardial injury. All manufacturers of cardiac troponin assays have claims approved by the FDA for use of their assay in the diagnosis of acute myocardial infarction (AMI) and have recommended a specific AMI cut-off concentration. For risk stratification purposes, it is also necessary to detect the presence of "minor" myocardial injury that occurs in patients with unstable angina. Lindahl and colleagues[35] and Antman and colleagues[36] showed that the use of cut-off concentrations for cTnT and cTnI that were lower than the recommended values for AMI diagnosis (0.18 versus 0.20 μg/L for cTnT, and 0.4 versus 1.5 μg/L for cTnI) identified additional individuals who were at risk for subsequent cardiovascular disease. Unfortunately, many manufacturers of cardiac troponin assays have not validated a low cut-off concentration for risk stratification purposes. As a result, different cut-off concentrations for the same assays have been used in various clinical studies (Table 3). Ideally, a laboratory should establish its own cut-off concentration for risk stratification purposes for the troponin assay being used. However, due to the need to determine patient outcomes at 4 to 6 weeks after initial presentation, it is difficult for most laboratories to conduct a risk stratification study. Therefore, one approach would be to use a cardiac troponin assay for which there is ample documentation of risk stratification. Even this may be difficult, as the instruments used in the original studies described in Table 3 (ES300, Stratus II, Opus) are obsolete and are being replaced with improved models. If a laboratory chooses to use a troponin assay that has not been validated in a clinical study, it is not possible to extrapolate an appropriate cut-off concentration for

Table 3

Assays and Cut-off Concentrations Used for Published Risk Stratification Studies

Reference	Assay[a]	Cut-off	Assay[b]	Cut-off
Luscher et al[28]	ES300	0.1	Opus	2.0
Hamm et al[29]	POC[c]	0.18	Status POC	1.5
Olatidoye et al[30]	ES300	0.1	Opus	1.6
Green et al[31]	ES300	0.14	Stratus II	1.9
Ottani et al[32]	ES300	0.1	Stratus II[d]	3.1

[a]Recommended cut-off for ES300: 0.1 μg/L; [b]Recommended cut-off for Opus: 2.0 μg/L, Stratus II: 1.5 μg/L; [c]POC = point-of-care testing device; [d]Premarket research assay.

risk stratification because of the lack of standardization between assays. In this situation, the most appropriate cut-off concentration for risk stratification would be to select the 95% value as recommended by the National Academy of Clinical Biochemistry (NACB)[37] or the 99% one-tailed value that is currently being considered by the American College of Cardiology.

Interassay Imprecision

Accurate discrimination between minor myocardial injury versus analytical noise requires assays that have high interassay precision at the low cut-off concentrations. The NACB has recommended a precision for cardiac troponin assays of 10% or less.[37] In the 1999 proficiency survey of the College of American Pathologists, this goal had not been met by any manufacturer when the survey sample with the lowest troponin concentration was compared, although some assays approached this limit (Table 4). For the purpose of risk stratification, cut-off concentrations for assays that are not precise must be increased to minimize the incidence of false-positive results due to analytical imprecision. This increase in cut-off concentration results in a decrease in the detection of minor myocardial injury cases and a degradation in the odds ratio for detection of undesirable outcomes (death and

Table 4

Results of the CAP Proficiency Survey for Cardiac Markers[a]

Manufacturer	No. Labs	Mean	S.D.	C.V.
Abbott Axsym	1363	5.30	0.85	16.1
Beckman Access	249	0.033	0.011	32.5
Biosite Triage[b]	57	0.416	0.130	31.5
Chiron ACS:180	190	0.854	0.122	14.3
Chiron Centaur	21	0.880	0.106	12.1
Dade Behring Magnum	12	1.42	0.49	34.5
Dade Behring Opus	24	1.15	0.41	35.3
Dade Behring Stratus CS	88	0.217	0.024	11.2
Dade Dimension	379	0.174	0.049	27.9
Stratus II	235	0.484	0.250	51.7
Technicon Immuno-1	108	1.15	0.13	11.0
Roche Elecsys	107	0.482	0.053	11.0

[a]1999 Cardiac Marker Survey Set CAR-C. [b]Note that this assay is not approved for use in serum, the matrix used in the survey material. CAP = College of American Pathologists; CV = coefficient of variation; SD = standard deviation.

AMI). Future-generation assays should be optimized for sensitivity and precision in lieu of a wide dynamic linear range. Concentrations exceeding the upper limit of linearity (ULL) can be appropriately diluted. Many immunoassay instruments can perform dilutions on line. Because the clinical value of very high cardiac troponin concentrations is questionable, it may be possible for a laboratory to report greater than 100 $\mu g/L$ (or the appropriate ULL concentration) and not dilute the sample. This practice will require acceptance by the ordering cardiologists and physicians.

Conclusion

There is now a large body of evidence that documents the notion that measurement for cardiac troponin provides additional clinical information over CK-MB for patients with acute coronary syndromes. Discussion on the use of cardiac troponin will have a prominent part in the next *American College of Cardiology/American Heart Association Guidelines for the Management of Patients with Unstable Angina*. Improvement in the quality of assays is the next critical step before there can be widespread acceptance of troponin. Unlike commercial assays for serum sodium, which all have high precision and produce the same answer, commercial assays for troponin have not matured to the point that all assays have the same performance. Clearly, some assays for troponin are superior to others. Unfortunately, a laboratory usually does not have an option to choose the best assay available. Often clinical pathologists and scientists must use instrumentation that is already in place in their laboratory and accept the quality of the troponin assay on that analyzer. Even the choice of instrumentation is increasingly being made on the basis of costs and membership to group buying for discounts, and not based on the quality of the equipment itself. Thus, it is the responsibility of manufacturers to improve the performance of their troponin assay and, in the case of cTnI, embrace the international efforts at assay standardization.

References

1. Dean KJ. Biochemistry and molecular biology of troponins I and T. In Wu AHB (ed): *Cardiac Markers*. Totowa, NJ: Humana Press; 1998:193–194.
2. Wu AHB, Feng YJ, Moore R, et al. Characterization of cardiac troponin subunit release into serum following acute myocardial infarction, and comparison of assays for troponin T and I. American Association for Clinical Chemistry Subcommittee on cTnI Standardization. *Clin Chem* 1998;44: 1198–1208.

3. Giuliani I, Bertinchant JP, Granier C, et al. Determination of cardiac troponin I forms in the blood of patients with acute myocardial infarction and patients receiving crystalloid or cold blood cardioplegia. *Clin Chem* 1999; 45:213–222.
4. Katrukha AG, Bereznikova AV, Esakova TV, et al. Troponin I is released in blood stream of patients with acute myocardial infarction not in free form but as a complex. *Clin Chem* 1997;43:1379–1385.
5. Wu AHB, Feng YJ. Biochemical differences between cTnT and cTnI and its significance for the diagnosis of acute coronary syndromes. *Eur Heart J* 1998;19(suppl N):25–29.
6. Katus HA, Looser S, Hallermayer K, et al. Development and in vitro characterization of a new immunoassay for cardiac troponin T. *Clin Chem* 1992; 38:386–393.
7. Mueller-Bardorff M, Halleymayer K, Schroeder A, et al. Improved troponin T ELISA specific for the cardiac troponin T isoform. Part I: Development, analytical and clinical validation of the assay. *Clin Chem* 1997;43:458–461.
8. Wu AHB. Laboratory and near patient testing for cardiac markers. *J Clin Ligand Assay* 1999;22:32–37.
9. Li D, Keffer J, Corry K, et al. Nonspecific elevation of troponin T levels in patients with chronic renal failure. *Clin Biochem* 1995;28:474–477.
10. Bhayana V, Gougoulias T, Cohoe S, Henderson R. Discordance between results for serum troponin T and troponin I in renal disease. *Clin Chem* 1995;41:312–317.
11. Kobayashi S, Tanaka M, Tamura N, et al. Serum cardiac troponin T in polymyositis/dermatomyositis. *Lancet* 1992;340:726.
12. Braun SL, Pongratz DE, Bialk P, et al. Discrepant results for cardiac troponin T and I in chronic myopathy, depending on instrument and assay generation. *Clin Chem* 1996;42:2039–2041.
13. Wu AHB, Feng YJ, Roper E, et al. Cardiac troponins T and I before and after renal transplantation. *Clin Chem* 1997;43:411–412.
14. Apple FS, Sharkey SW, Hoeft P, et al. A 1-year outcomes analysis. *Am J Kidney Dis* 1997;29:399–403.
15. Ricchiuti V, Voss EM, Ney A, et al. Cardiac troponin T isoforms expressed in renal diseased skeletal muscle will not cause false-positive results by the second generation cardiac troponin T assay by Boehringer Mannheim. *Clin Chem* 1998;44:1919–1924.
16. Haller C, Zehelein J, Remppis A, et al. Cardiac troponin T in patients with end-stage renal disease: Absence of expression in truncal skeletal muscle. *Clin Chem* 1998;44:930–938.
17. Porter GA, Norton T, Bennett WB. Troponin T, a predictor of death in chronic haemodialysis patients. *Eur Heart J* 1998;19(suppl):N34-N37.
18. Mockel M, Schindler R, Knorr L, et al. Prognostic value of cardiac troponin T and I elevations in renal disease patients without acute coronary syndromes: A 9-month outcome analysis. *Nephrol Dial Transplant* 1999;14: 1489–1495.
19. Musso P, Cox I, Vidano E, et al. Cardiac troponin elevations in chronic renal failure: Prevalence and clinical significance. *Clin Biochem* 1999;32:125–130.
20. Van Lente F, McErlean ES, DeLuca SA, et al. Ability of troponins to predict adverse outcomes in patients with renal insufficiency and suspected acute coronary syndromes: A case-matched study. *J Am Coll Cardiol* 1999;33: 471–478.
21. Roberts WL, Calcote CB, De BK, et al. Prevention of analytical false-positive

increases of cardiac troponin I on the Stratus II analyzer. *Clin Chem* 1997; 43:860–861.
22. Volk A, Hardy R, Robinson CA, Konrad RJ. False-positive cardiac troponin I results. Two case reports. *Lab Med* 1999;30:610–612.
23. Fitzmaurice TF, Brown C, Rifai N, et al. False increase of cardiac troponin I with heterophilic antibodies. *Clin Chem* 1998;44:2212–2214.
24. Bohner J, von Pape KW, Hannes W, Stegmann T. False-negative immunoassay results for cardiac troponin I probably due to circulation troponin I autoantibodies. *Clin Chem* 1996;42:2046–2047.
25. Kaplan IV, Levinson SS. When is a heterophile antibody not a heterophile antibody? When it is an antibody against a specific immunogen. *Clin Chem* 1999;45:616–618.
26. Hamm CW, Ravkilde J, Gerhardt W, et al. The prognostic value of serum troponin T in unstable angina. *N Engl J Med* 1992;327:146–150.
27. Ohman EM, Armstrong PW, Christenson RH, et al. Cardiac troponin T levels for risk stratification with admission cardiac troponin T levels in acute myocardial ischemia. The GUSTO IIa Investigators. *N Engl J Med* 1996;335:1333–1341.
28. Luscher MS, Thygesen K, Ravkilde J, Heickendorff L, for the TRIM Study Group. Applicability of cardiac troponin T and I for early risk stratification in unstable coronary artery disease. *Circulation* 1997;96:2578–2585.
29. Hamm CW, Goldmann BU, Heeschen C, et al. Emergency room triage of patients with acute chest pain by means of rapid testing for cardiac troponin T or troponin I. *N Engl J Med* 1997;337:1648–1653.
30. Olatidoye AG, Wu AHB, Feng YJ, Waters D. Prognostic role of troponin T versus troponin I in unstable angina pectoris for cardiac events with meta-analysis comparing published studies. *Am J Cardiol* 1998;81: 1405–1410.
31. Green GB, Li DJ, Bessman ES, et al. The prognostic significance of troponin I and troponin T. *Acad Emerg Med* 1998;5:758–767.
32. Ottani F, Galvani M, Ferrini D, et al. Direct comparison of early elevations of cardiac troponin T and I in patients with clinical unstable angina. *Am Heart J* 1999;137:284–291.
33. Christenson RH, Duh SH, Newby LK, et al, for the GUSTO-IIa Investigators. Cardiac troponin T and cardiac troponin I: Relative values in short-term risk stratification of patients with acute coronary syndromes. *Clin Chem* 1998;44:494–501.
34. Newby LK, Christenson RH, Ohman EM, et al. Value of serial troponin T measures for early and late risk stratification in patients with acute coronary syndromes. The GUSTO IIa Investigators. *Circulation* 1998;98: 1853–1859.
35. Lindahl B, Venge P, Wallentin L, for the FRISC Study Group. Relation between troponin T and the risk of subsequent cardiac events in unstable coronary artery disease. *Circulation* 1996;93:1651–1657.
36. Antman EM, Tanasijevic MJ, Thompson B, et al. Cardiac-specific troponin I levels to predict the risk of mortality in patients with acute coronary syndromes. *N Engl J Med* 1996;335:1342–1349.
37. Wu AHB, Apple FS, Gibler WB, et al. National Academy of Clinical Biochemistry Standards of Laboratory Practice: Recommendations for use of cardiac markers in coronary artery diseases. *Clin Chem* 1999;45:1104–1121.

Chapter 2

Prepare to Meet Your Markers: *Making the Most out of Troponin I Degradation*

Jason L. McDonough, BSc(H), Ralf Labugger, MSc, and Jennifer E. Van Eyk, PhD

Every cardiac disease, whether due to acute injury such as myocardial infarction or to chronic conditions such as heart failure or dilated cardiomyopathy, produces changes in the protein profile of the cardiac myocyte.[1–3] These disease-induced protein modifications may be the result of post-translational modifications (including proteolysis, covalent cross-linking, phosphorylation, oxidation, and others) or of modified gene expression (including upregulation, downregulation, and isoform switching). The overall protein profile of the myocyte, also called the proteome, is thus a spectrum of protein modifications resulting from the numerous overlapping disease processes in any given patient.

Many ischemic conditions including acute coronary syndromes and chronic diseases such as heart failure result in the release of intracellular proteins into the blood. Currently used biomarkers that are sensitive to myocardial injury are based on the detection of these intracellular proteins, including troponin I (cTnI), troponin T (cTnT), myoglobin, and the creatine kinases (CK and CK-MB).

Cardiac TnI and cTnT are unique among these cardiac biomarkers because they are an essential part of the contractile apparatus. Therefore, they may provide information about the 'functional consequences' of myocardial injury—in other words, the impact on the contractility of the remaining viable myocardium. This is especially pertinent since cTnI is specifically and selectively modified in the tissue of diseased hearts.[4–7] Importantly, cTnI undergoes progressive post-translational modification with ischemia/reperfusion injury, including proteoly-

From: Adams JE III, Apple FS, Jaffe AS, Wu AHB (eds). *Markers in Cardiology: Current and Future Clinical Applications.* Armonk, NY: Futura Publishing Company, Inc.; © 2001.

sis,[8–11] formation of covalent cross-links,[9,10] and phosphorylation.[9–12] These various disease-induced cTnI modification products are present in the myocytes and, when released into the blood, will form a profile that reflects the state of the heart at the time of injury. The serum cTnI profile should reflect the progression of the disease, including the severity and the time from onset of injury, as well as the functional status of the remaining viable myocardium. This information could affect treatment of patients and enable physicians to provide more precise and individualized long-term prognosis and management.

The Functional Importance of cTnI

Cardiac muscle contraction occurs through the tightly regulated interactions of troponin-tropomyosin (Fig. 1).[13–17] Troponin consists of three proteins: TnT, which interacts with tropomyosin; TnI, the inhibitory protein; and troponin C (TnC), which binds Ca^{2+}. There are extensive interactions between cTnI and cTnC, cTnT, and actin-tropomyosin,

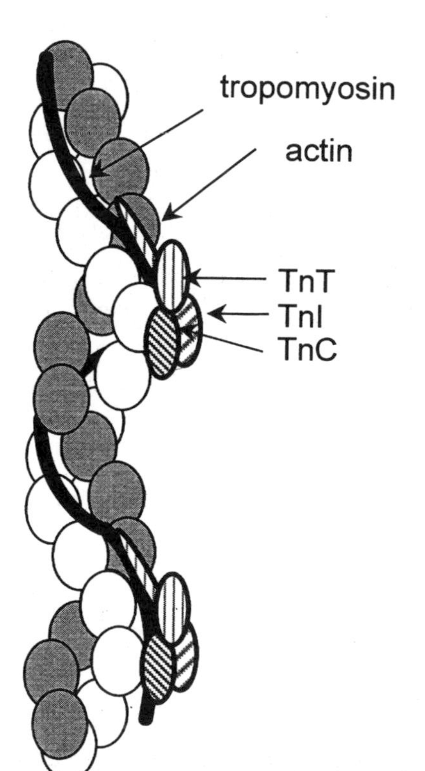

Figure 1. The actin-tropomyosin filament. A schema of the actin-tropomyosin filament of cardiac muscle, with the trimeric troponin regulatory complex. Troponin I (cTnI) switches from binding sites on actin to troponin C (TnC) in the presence of calcium (refer to text for details).

some of which are Ca-dependent and others of which are Ca-independent. Troponin-tropomyosin binds the actin filament and controls actin-myosin interactions through the intricate interplay of both steric and allosteric mechanisms.[18,19]

In diastole, or the 'off' state, myosin is sterically prevented from interacting with actin through the interaction of cTnI with actin and tropomyosin. During systole, or the 'on' state, the flood of Ca^{2+} from the sarcoplasmic reticulum results in the Ca-dependent movement of the troponin-tropomyosin complex, and the interaction between myosin and actin. This in turn promotes the recruitment of other myosin heads to the actin filament (a mechanism called allosteric activation). Ca^{2+} binding to cTnC increases the affinity between cTnI and cTnC while weakening interactions between actin and cTnI and interactions between the C-terminus of cTnT and tropomyosin. Meanwhile, the N-terminus of cTnT remains bound to tropomyosin and is proposed to hold troponin-tropomyosin complex on the actin filament irrespective of the intracellular Ca^{2+} concentration. Cardiac TnI regulates muscle contraction by switching from its binding site on actin-tropomyosin to cTnC in the presence of Ca^{2+}. The role of cTnT has been simplistically ascribed to its interaction with tropomyosin, acting as an anchoring protein, holding or positioning troponin on the actin-tropomyosin filament. However, the actual role of cTnT is more complex since it modulates the myofilament response to Ca^{2+}, enhancing the interaction between actin and myosin.[20,21]

The myofilament system is also highly cooperative. This means that local activation of a single troponin in response to Ca^{2+} can be propagated along the actin filament affecting adjacent troponin-tropomyosin units. Myofilament cooperativity can help explain why low expression levels of genetically altered cTnI[22] or cTnT[23,24] result in large, or even lethal, contractile changes in transgenic animals. Given the structural complexity of the troponin-tropomyosin-actin assembly, as well as the allosteric and cooperative components of the mechanism, one can appreciate that the cardiac myofilaments constitute a finely tuned system for the regulation of force production. As a corollary, subtle modifications to the myofilament proteins, including disease-induced modifications, should alter cardiac contraction.

Disease-Induced Protein Modifications in the Myocardium

Identification of disease-induced protein changes in the myocardium is critical to elucidate the molecular mechanisms of the pathologic condition and, since these proteins may eventually be released from the cell into the blood, it forms the basis for a diagnosis. Modification to many myofilament proteins, including proteolysis and cross-linking of cTnI[8–10] and cTnT,[10,25–27] and alterations of α-actinin,[9,27] myosin light chain,[9] desmin,[27] and spectrin[27] have been detected in the myocardium in a number of animal models subjected to ischemia of variable severity. The largest body of work on ischemia-induced modification of cTnI comes from the analysis of isolated Langendorff-perfused rat hearts. Cardiac TnI was specifically and selectively degraded in hearts that experienced mild ischemia/reperfusion (15 or 20 minutes of ischemia followed by 20 or 45 minutes of reperfusion),[8,9] in which there was no detectable necrosis[9]—a condition mimicking myocardial stunning. The cTnI degradation correlates with a decrease in the maximum force of contraction of intact trabeculae[28] or skinned muscle fiber bundles.[8,9]

Further analysis demonstrated that a spectrum of cTnI modification occurs with increasing severity of the ischemia/reperfusion injury[9,10] (Fig. 2). With mild ischemia, cTnI was degraded to a single 22-

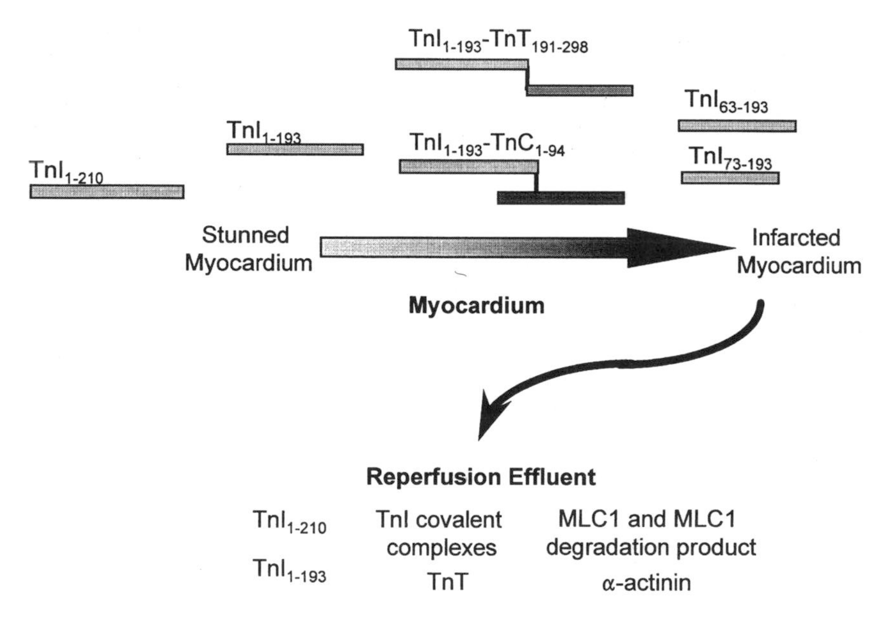

Figure 2. The troponin I (TnI) modification products found in myocardium and reperfusion effluent from isolated perfused rat hearts. TnI is progressively and selectively modified with the increasing severity of ischemia/reperfusion injury in isolated rat hearts.[9,10] As injury progresses from myocardial stunning (with no cell death) to infarction (with necrosis), cTnI becomes more extensively modified. Severe ischemia results in the release of cellular proteins into the reperfusion effluent, including cTnI modification products (refer to text for details).

kd product,[8–10] the result of a loss of 17 amino acid residues from the C-terminus producing the cTnI fragment, $cTnI_{1-193}$.[10] Other early changes included the formation of binary nondisulfide covalent complexes between the troponin subunits,[9,10] loss of α-actinin,[9] and phosphorylation of cTnI.[9,10,12] Increased severity of ischemia (60 minutes of ischemia followed by 45 minutes of reperfusion) resulted in more extensive degradation of cTnI from the N-terminus, producing two additional cTnI fragments (16 kd and 15 kd, proposed to be cTnI residues 63–193 and 73–193, respectively).[10]

The in vivo functional consequences of cTnI degradation were explored by generating a line of transgenic mice that expressed the degradation product $cTnI_{1-193}$ (constituting 9% to 20% of total cTnI, a level similar to that in stunned rat myocardium).[22] These mice exhibited ventricular dilation and diminished contractility, recapitulating the phenotype of myocardial stunning. Previous in vitro work using truncated forms of recombinant cTnI demonstrated that the extreme C-terminal region (amino acid residues 188–210) modulates calcium sensitivity.[29,30] Similar in vitro studies of the functional consequences of the exact N-terminal degradation observed with severe ischemia have yet to be performed. However, within the first 73 amino acids of cTnI lie the phosphorylation sites for protein kinase A, as well as extensive cTnC and cTnT binding sites (for example[31–34]). Clearly, the exact site of cTnI proteolysis will dictate the extent of contractile dysfunction, depending on which interactions are affected.

From a diagnostic perspective, the most useful protein changes are those that are progressive and are key to the phenotypic expression of the disease. This is especially important if one is attempting to differentiate between a "generic" marker of myocardial damage and a marker that might indicate "functional" prognosis. The work on ischemia/reperfusion in isolated rat hearts indicates that cTnI and its modification products are potential candidates for such a marker.

In our most recent work, modification of cTnI was observed in the myocardium of coronary artery bypass patients, even before the application of the cross-clamp.[22,35] The profiles of cTnI modification observed in left ventricular myocardial biopsies from bypass patients were similar to those in ischemia/reperfused isolated rat hearts. There was a limited number of degradation products of cTnI, and the number and the quantity of each product varied between patients.[35] These differences in the cTnI modification profiles might reflect variations in the state of disease in the myocardium.

Given that cTnI modification products are present in human post-ischemic myocardium, it seems probable that cTnI modification products are eventually released into the serum. This is supported by the detection of cTnI degradation products, cTnI-containing covalent complexes, α-actinin, and many other myofilament proteins in the reperfusion effluent from severely ischemic isolated rat hearts[9] (Fig. 2). In serum, the detection of cTnI modification products originating from the myocardium is much more complex.

Detection of cTnI in Serum

The parameters that affect the quantity and forms of intracellular proteins (such as cTnI) that are present in the serum include any disease-induced processing of the protein in the myocardium and any processing in the blood once released from the myocardium. The appearance for each protein (or its modified products) in the blood will occur at different times depending on its molecular mass, its affinity for other proteins, and its localization in the cell.[36] In addition, their susceptibility to proteolysis and filtration by the renal and lymphatic systems will affect their disappearance from the blood. Therefore, different protein biomarkers will appear at different times after the onset of an ischemic event, contributing (together with specificity and sensitivity) to the merits and shortcomings of the several commercial diagnostic tests. For example, as with CK and CK-MB, the troponins are detectable approximately 6 hours after onset of a myocardial infarction, but stay elevated for up to 3 weeks.[37,38] Therefore, troponins can also serve as markers for distant ischemic events. Their higher sensitivity (compared with myoglobin, CK and CK-MB) is useful in the diagnosis of heart failure,[39] unstable angina (in terms of prognosis for future cardiac events),[40] and minor myocardial damage[38] in cases with nonelevated CK-MB levels. Cardiac TnI and cTnT may also allow for a more complete characterization of heart failure.

A drawback of the existing cTnI serum diagnostics is the lack of standardization. Tate et al.[41] recently showed, comparing four diagnostic tests, up to a 60-fold difference in absolute values for patient samples. Interestingly, all of the cTnI tests identified the positive samples, and no negative samples were detected as false positive.[41] How is it possible that there are enormous differences in absolute values, when each test is internally consistent? Although post-translational modification of cTnI is often blamed,[42,43] the problem is not that cTnI modification occurs, but rather it is in the selection of the anti-cTnI antibodies used. What is required are anti-cTnI antibodies that are capable of recognizing all of the possible cTnI modification products that may be present in the blood. One could exploit the fact that certain cTnI modification products may be present in a patient's blood at specific times during the development of a disease. Raising antibodies that recognize these cTnI modification products may allow for further clinical stratification. Certain degradation products of cTnI have a distinct pattern of release over time.[44] This may lead to tests that provide more precise information about the severity of damage, the time of onset, or even the type of disease.

The main limitation is the lack of knowledge of the exact forms of cTnI circulating in the patient's blood.[45,46] Using a modified sodium dodecyl sulfate-polyacrylamide gel electrophoresis protocol, followed by immunoblot analysis, we have recently identified intact cTnI as the predominant form found in serum after acute myocardial infarction.[44] In addition to intact cTnI, cTnI degradation products were also present in a subset of patients. Unlike previous reports,[42,47] when we spiked serum with very low, pathologically relevant quantities of cTnI, only minimal proteolysis occurred.[44] Therefore, some of the cTnI modification products observed in the serum most likely originated in the myocardium. These forms of cTnI display a characteristic rising and falling pattern after the onset of symptoms, producing a continuum of changing cTnI profiles. This new method for observing and detecting cTnI and its modification products in the serum should allow for the optimization and standardization of the current cTnI diagnostic kits. We also hope that the appearance of specific cTnI modification products in the serum can be used to stratify patients with acute coronary syndrome and other heart diseases, thereby improving treatment.

References

1. Pleissner KP, Soding P, Sander S, et al. Dilated cardiomyopathy-associated proteins and their presentation in a WWW-accessible two-dimensional gel protein database. *Electrophoresis* 1997;18:802–808.
2. Weekes J, Wheeler CH, Yan JX, et al. Bovine dilated cardiomyopathy: Proteomic analysis of an animal model of human dilated cardiomyopathy. *Electrophoresis* 1999;20:898–906.
3. Dunn MJ. Studying heart disease using the proteomic approach. *Drug Discov Today* 2000;5:76–84.
4. Bolli R, Marban E. Molecular and cellular mechanisms of myocardial stunning. *Physiol Rev* 1999;79:609–634.
5. Lamers JM. Preconditioning and limitation of stunning: One step closer to the protected protein(s)? *Cardiovasc Res* 1999;42:571–575.
6. Solaro RJ. Troponin I, stunning, hypertrophy, and failure of the heart. *Circ Res* 1999;84:122–124.
7. Foster DB, Van Eyk JE. In search of the proteins that cause myocardial stunning. *Circ Res* 1999;85:470–472.
8. Gao WD, Atar D, Liu Y, et al. Role of troponin I proteolysis in the pathogenesis of stunned myocardium. *Circ Res* 1997;80:393–399.
9. Van Eyk JE, Powers F, Law W, et al. Breakdown and release of myofilament proteins during ischemia and ischemia/reperfusion in rat hearts: Identifica-

tion of degradation products and effects on the pCa-force relation. *Circ Res* 1998;82:261–271.
10. McDonough JL, Arrell DK, Van Eyk JE. Troponin I degradation and covalent complex formations accompanies myocardial ischemia/reperfusion injury. *Circ Res* 1999;84:9–20.
11. Zakhary DR, Moravee CS, Stewart RW, et al. Protein kinase A (PKA)-dependent troponin I phosphorylation and PKA regulatory subunits are decreased in human dilated cardiomyopathy. *Circulation* 1999;99:505–510.
12. Van Eyk JE, Organ LR, Buscemi N, et al. Cardiac disease-induced post-translational modifications of troponin I: Differential proteolysis, phosphorylation and covalent complex formation. *Biophys J* 2000;78:107A.
13. Filatov VL, Katrukha AG, Bulargina TV, et al. Troponin: Structure, properties, and mechanism of functioning. *Biochemistry (Moscow)* 1999;64:969–985.
14. Perry SV. Troponin I: Inhibitor or facilitator. *Mol Cell Biochem* 1999;190: 9–32.
15. Solaro RJ, Rarick HM. Troponin and tropomyosin: Proteins that switch on and tune in the activity of cardiac myofilaments. *Circ Res* 1998;83:471–480.
16. Solaro RJ, Van Eyk JE. Altered interactions among thin filament proteins modulate cardiac function. *J Mol Cell Cardiol* 1996;28:217–230.
17. Van Eyk JE, Hodges RS. The use of synthetic peptides to unravel the mechanism of muscle regulation. *Methods* 1993;5:264–280.
18. Lehrer SS. The regulatory switch of the muscle thin filament: Ca^{2+} or myosin heads? *J Muscle Res Cell Motil* 1994;15:232–236.
19. Lehrer SS, Geeves MA. The muscle thin filament as a classical cooperative/allosteric regulatory system. *J Mol Biol* 1998;277:1081–1089.
20. Tobacman LS. Thin filament-mediated regulation of cardiac contraction. *Annu Rev Physiol* 1996;58:447–481.
21. Chalovich JM. Actin mediated regulation of muscle contraction. *Pharmacol Ther* 1992;55:95–148.
22. Murphy AM, Kogler H, Georgakopolous D, et al. Transgenic mouse model of stunned myocardium. *Science* 2000;287:488–491.
23. Tardiff JC, Factor SM, Tompkins BD, et al. A truncated cardiac troponin T molecule in transgenic mice suggests multiple cellular mechanisms for familial hypertrophic cardiomyopathy. *J Clin Invest* 1998;101:2800–2811.
24. Tardiff JC, Hewett TE, Palmer BM, et al. Cardiac troponin T mutations result in allele-specific phenotypes in a mouse model for hypertrophic cardiomyopathy. *J Clin Invest* 1999;104:469–481.
25. Gorza L, Menabo R, Di Lisa F, et al. Troponin T cross-linking in human apoptotic cardiomyocytes. *Am J Pathol* 1997;150:2087–2097.
26. Gorza L, Menabo R, Vitadello M, et al. Cardiomyocyte troponin T immunoreactivity is modified by cross-linking resulting from intracellular calcium overload. *Circulation* 1996;93:1896–1904.
27. Matsumura Y, Saeki E, Inoue M, et al. Inhomogeneous disappearance of myofilament-related cytoskeletal proteins in stunned myocardium of guinea pig. *Circ Res* 1996;79:447–454.
28. Gao WD, Atar D, Backx PH, et al. Relationship between intracellular calcium and contractile force in stunned myocardium. Direct evidence for decreased myofilament Ca^{2+} responsiveness and altered diastolic function in intact ventricular muscle. *Circ Res* 1995;76:1036–1048.
29. Rarick HM, Tu XH, Solaro RJ, et al. The C-terminus of cardiac troponin I is essential for full inhibitory activity and Ca-sensitivity of rat myofibrils. *J Biol Chem* 1997;272:26887–26892.

30. Ramos CH. Mapping subdomains in the C-terminal region of troponin I involved in its binding to troponin C and to thin filament. *J Biol Chem* 1999; 274:18189–18195.
31. Rarick HM, Tang HP, Guo XD, et al. Interactions at the NH2-terminal interface of cardiac troponin I modulate myofilament activation. *J Mol Cell Cardiol* 1999;31:363–375.
32. Jaquet K, Lohmann K, Czisch M, et al. A model for the function of the bisphosphorylated heart-specific troponin-I N-terminus. *J Muscle Res Cell Motil* 1998;19:647–659.
33. Szczesna D, Zhang R, Zhao J, et al. The role of the NH(2)- and COOH-terminal domains of the inhibitory region of troponin I in the regulation of skeletal muscle contraction. *J Biol Chem* 1999;274:29536–29542.
34. Van Eyk JE, Thomas LT, Tripet B, et al. Distinct regions of troponin I regulate Ca^{2+}-dependent activation and Ca^{2+} sensitivity of the acto-S1-TM ATPase activity of the thin filament. *J Biol Chem* 1997;272:10529–10537.
35. McDonough JL, Ropchan G, Atar D, et al. Biochemistry in the OR: cTnI modification in bypass surgery. *Circulation* 1999;100:I766. Abstract.
36. Adams JE. Clinical application of markers of cardiac injury: Basic concepts and new considerations. *Clin Chim Acta* 1999;284:127–134.
37. Storrow AB, Gibler WB. The role of cardiac markers in the emergency department. *Clin Chim Acta* 1999;284:187–196.
38. Hudson MP, Cristenson RH, Newby LK, et al. Cardiac markers: Point of care testing. *Clin Chim Acta* 1999;284:223–237.
39. Missov ED, Calzolari C, Pau B. Circulating cardiac troponin I in severe congestive heart failure. *Circulation* 1997;96:2953–2958.
40. Galvani M, Ottani F, Ferrini D. Prognostic influence of elevated values of cardiac troponin I in patients with unstable angina. *Circulation* 1997;95: 2053–2059.
41. Tate JR, Heathcote D, Rayfield J, et al. The lack of standardization of cardiac troponin I assay systems. *Clin Chim Acta* 1999;284:141–149.
42. Shi Q, Ling M, Zhang X, et al. Degradation of cardiac troponin I in serum complicates comparisons of cardiac troponin I assays. *Clin Chem* 1999;45: 1018–1025.
43. Katrukha AG, Bereznikova AV, Filatov VL, et al. Degradation of cardiac troponin I: Implication for reliable immunodetection. *Clin Chem* 1998;44: 2433–2440.
44. Labugger R, Organ L, Collier C, et al. Extensive troponin I and T modification detected in serum from patients with acute myocardial infarction. *Circulation* 2000;102:1221–1226.
45. Katus HA, Rempis A, Neumann FJ, et al. Diagnostic efficiency of troponin T measurements in acute myocardial infarction. *Circulation* 1991;83:902–912.
46. Wu AH, Feng YJ, Moore R, et al. Characterization of cardiac troponin subunit release into serum after acute myocardial infarction and comparison of assays for troponin T and I. American Association for Clinical Chemistry Subcommittee on cTnI Standardization. *Clin Chem* 1998;44:1198–1208.
47. Morjana NA. Degradation of human cardiac troponin I after myocardial infarction. *Biotechnol Appl Biochem* 1998;8:105–111.

Chapter 3

Functional Sensitivity of Cardiac Troponin Assays and its Implications for Risk Stratification for Patients with Acute Coronary Syndromes

Kiang-Teck J. Yeo, PhD, Kelly S. Quinn-Hall, MT, Stephanie W. Bateman, BA, George A. Fischer, PhD, Stacey Wieczorek, PhD, and Alan H.B. Wu, PhD

Introduction

Cardiac troponins T (cTnT) and I (cTnI) have high specificity for cardiac injury because the cardiac-specific antibodies used in these assays do not cross-react with the corresponding skeletal muscle isoforms of troponin.[1] Assays for troponin are especially sensitive tools for detecting low concentrations of protein release in patients suffering minor myocardial injury, because the heart has a very high tissue content of troponin T (10.8 mg/g wet weight) and I (6 mg/g) as compared with the protein content of other serum markers such as creatine kinase MB isoenzyme (1.4 mg/g).[2] In addition, because of the high antibody specificity for troponin and the absence of significant troponin concentrations in sera of normal individuals, very low cut-off concentrations can be used to maximize the detection of low abnormal troponin levels.[3]

Detection of minor injury is clinically important for prognostic assessments in patients with unstable angina. Clinical studies have shown that an abnormal concentration of cTnT or cTnI is associated

From: Adams JE III, Apple FS, Jaffe AS, Wu AHB (eds). *Markers in Cardiology: Current and Future Clinical Applications.* Armonk, NY: Futura Publishing Company, Inc.; © 2001.

with a four- to fivefold increase in risk for acute myocardial infarction (AMI) and death within a 6-month follow-up period.[4,5] Thus, cardiac troponin has been proposed to be useful in risk stratification for patients who have serum concentrations within the "indeterminate" zone, ie, between the upper reference limit (URL) and the AMI cut-off points.[6] However, clinicians are wary of using low abnormal results for risk stratification because of anecdotal reports of false-positive interferences resulting from both the presence of heterophile and human anti-mouse antibodies[7] and the presence of fibrin strands.[8] With the introduction of second- and third-generation troponin assays, many of these issues have been, or are in the process of being, resolved.[9] A continuing problem for troponin assays, however, is assay imprecision in the low concentration range. For most clinicians and laboratorians, the assessment of functional sensitivity for troponin is a particularly difficult issue to understand for cTnI because there is no standardization of results between commercial assays,[10] and package inserts do not uniformly address the issues of cut-off concentrations and assay sensitivities. Thus, this study was necessary to independently determine the functional sensitivities of five commercial cTnI assays and one cTnT assay using a single protocol and the same pooled plasma study materials.

Patient plasma specimens were selected and pooled to target four different levels of cTnI spanning the low normal to AMI cut-off ranges (based on the Abbott AxSYM method). Each pool was filtered and centrifuged at 1000 xg to remove particulate debris and fibrin strands. The four pools were stored as aliquots at −20°C until ready for assay. Over a period of 8 to 10 weeks, these aliquots were measured by several commercial cardiac troponin assays to determine the mean and coefficient of variation (CV) at the four troponin levels according to each manufacturer's procedure. The cTnI assays were performed on the following systems: Dade Behring Dimension RxL (Deerfield, IL), Bayer Centaur (Tarrytown, NY), Abbott AxSYM (Abbott Park, IL), Johnson & Johnson Vitros ECi (Rochester, NY), and DPC Turbo Immulite (Diagnostics Products Corporation, Los Angeles, CA). Cardiac TnT was performed on the Roche Elecsys 2010 system (Roche Diagnostics, Indianapolis, IN).

The mean and standard deviation values were calculated for each patient pool based on between-day runs for each cardiac troponin assay. Precision profiles were constructed, and the functional sensitivity of each assay was determined from the interpolation of the 20% CV line with the respective precision profile curve (Fig. 1). The results for each assay are summarized in Table 1, with the manufacturer's claimed minimal detection limit (MDL), URL for normal individuals, and upper cut-off limit for AMI. We considered the indeterminate zone the implied values of cardiac troponin between the URL and the AMI cut-off. The Dimension RxL, Immulite, AxSYM, and Elecsys troponin

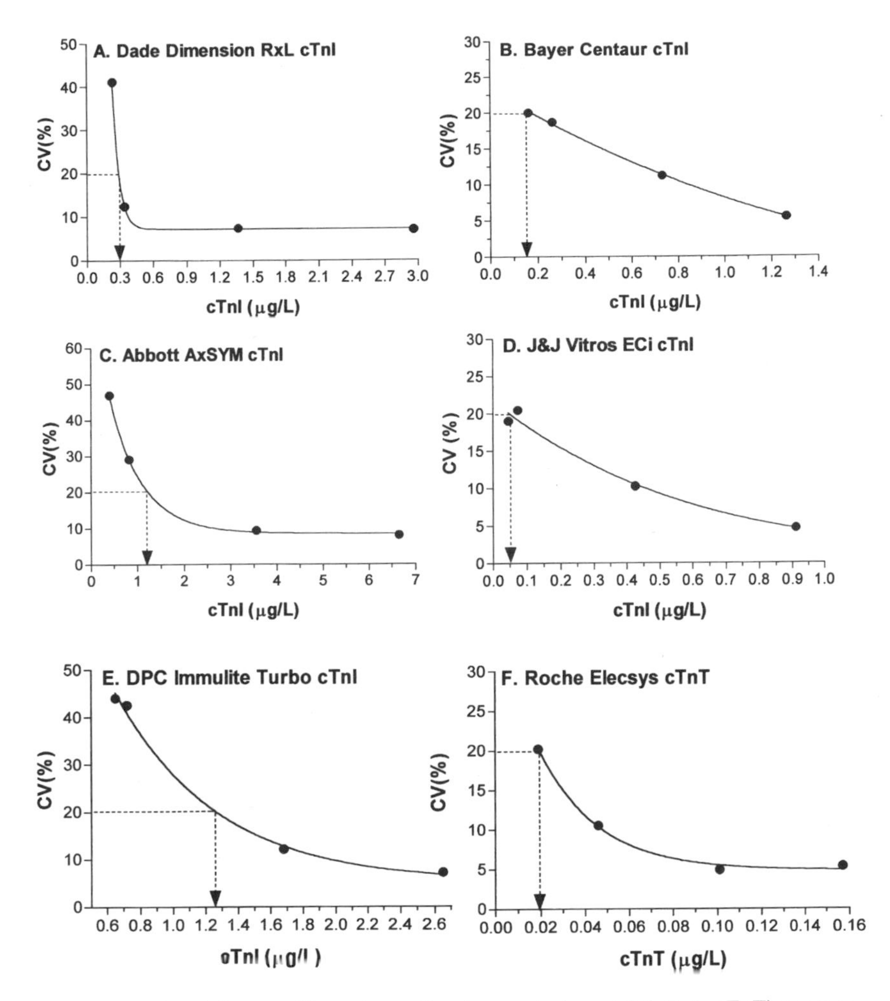

Figure 1. Precision profiles of cardiac troponin I and T (cTnI, cTnT) assays.

assays have functional sensitivities much higher than their respective URL cut-offs, whereas the Centaur assay has a functional sensitivity of 0.16 μg/L, which is just slightly above the URL cut-off of 0.15 μg/L. In contrast, the Vitros ECi cTnI assay has a functional sensitivity of 0.05 μg/L, which is below the URL cut-off point of 0.10 μg/L.

There is now substantial evidence that an abnormal cTnT or cTnI concentration is associated with increased risk for adverse cardiac events in patients with ACSs.[11–17] For example, a recent meta-analysis of 14 trials showed a hazard ratio of 2.7 for cTnT and 4.2 for cTnI for risk of AMI or death during a 30-day median follow-up period.[16] In addition, there is recent evidence that elevated cardiac troponin

Table 1

Functional Sensitivity of Cardiac Troponin Assays

Troponin Assay	Functional Sensitivity (μg/L)	MDL (μg/L)	URL Cut-off* (μg/L)	AMI Cut-off† (μg/L)	Indeterminate Zone (μg/L)
Dimension RxL cTnI	0.30	0.04	<0.06	≥0.60	0.06–0.59
Centaur cTnI	0.16	0.15	<0.15	≥1.50	0.15–1.40
AxSYM cTnI	1.25	0.30	<0.50	≥3.50‡	0.50–3.40
Vitros ECi cTnI	0.05	0.02	<0.10	≥1.00	0.10–1.00
Immulite Turbo cTnI	1.26	0.15	<1.00	≥1.00	not stated
Roche Elecsys cTnT	0.02	0.01	<0.01	≥0.10	0.01–0.09

Based on manufacturers' cited cut-offs for normal* and AMI.† ‡ Based on in-house determined AMI cut-off. AMI = acute myocardial infarction; MDL = minimal detection limit; URL = upper reference limit.

levels can be used as a criterion for selection of ACS patients who will benefit most from antithrombotic regimens. The FRISC study by Lindahl et al.[18] showed that patients with cTnT ≥0.1 μg/L that were treated with long-term dalteparin had a rate of adverse incidences (at 40 days) of 7.4% compared with 14.2% in the placebo group. Thus, using a cTnT cut-off of ≥0.1 μg/L in the first day after admission can identify a subgroup of patients with unstable coronary artery disease that will optimally benefit from prolonged dalteparin therapy. Hamm et al.,[19] in a substudy of the CAPTURE trial, showed that cTnT-positive patients benefited most from abciximab antithrombotic therapy during angioplasty. A 2.5-fold reduction in death or nonfatal MI (30-day follow-up) was observed in cTnT-positive patients who were treated with abciximab; in contrast, cTnT-negative patients showed no difference in 30-day outcomes when treated with abciximab compared with placebo. More recently, in the PRISM study, Heeschen et al.[20] showed that when using a cut-off of 1.0 μg/L for cTnI by the AxSYM method, or 0.1 μg/L for cTnT by the Elecsys 2010 method, elevated levels of either marker can be used reliably to identify high-risk patients with ACS who will benefit optimally from tirofiban (a platelet receptor antagonist) therapy.

Despite these large studies on the utility of troponin levels for both early risk stratification and selection of high-risk ACS patients for antithrombotic regimens, several factors have caused clinicians to be skeptically cautious concerning the use of low abnormal levels for these purposes. High analytical imprecision of cardiac troponin assays (especially for some cTnI assays) at the low dynamic range is a major factor that affects the reliability of using troponin values within the indeterminate zone (ie, abnormal values between the URL and AMI cut-off points). For most immunoassays, the manufacturer commonly cites the

MDL, the lowest concentration of analyte that is statistically different from zero. MDL is typically determined from an intra-assay determination of a zero calibrator,[21] and is not sufficiently robust to account for inter-assay variations. A further difficulty is that a manufacturer may use nonserum materials in determining the MDL, which may not translate adequately for serum or plasma determinations. Thus a manufacturer's stated MDL is rarely duplicated in clinical practice, and has been considered clinically irrelevant for immunoassays such as thyroid-stimulating hormone (TSH).[22]

Spencer et al.[23] have proposed using functional sensitivity, instead of the MDL, as a clinically relevant estimate of the lowest reporting limit for TSH. This parameter incorporates the imprecision associated with inter-assay variables (different calibrations, different reagent lots) and biological variables. By this method, inter-assay precision profiles are generated from several appropriate patient serum pools over a realistic time period, and the dose that corresponds to a 20% CV of the assay is then defined as the functional sensitivity. Current "ultrasensitive" TSH assays use this concept of functional sensitivity to differentiate between assays of greatly improved lower detection limit (third generation) and others that are less sensitive (second generation). TSH assays with lower functional sensitivities have been shown to result in fewer misclassifications of low TSH as normal.

In our study, we determined the precision profiles, over a clinically relevant time period, of five common cTnI assays and one common cTnT assay in order to measure the functional sensitivity of each method. Our results (Table 1) confirmed that for all the troponin assays, the functional sensitivity is consistently above the assay's stated MDL. The Dade Behring Dimension RxL, Abbott AxSYM, DPC Turbo Immulite cTnI, and Roche Elecsys cTnT assays showed functional sensitivity values that exceeded their respective URL cut-offs. This compromises the ability to reliably use low abnormal values within the indeterminate range for risk stratification purposes. The proposed Abbott AxSYM cTnI cut-off of greater than 1.0 μg/L[20] as a criterion for selection of ACS patients for antithrombotic therapy may not be appropriate in view of a functional sensitivity of 1.25 μg/L. The DPC Immulite cTnI functional sensitivity of 1.26 μg/L exceeded the AMI cut-off, indicating that numerous false-positive cTnI results would arise from using the published cut-off of 1.0 μg/L. In contrast, the functional sensitivity of the J&J Vitros ECi cTnI assay is well below the indeterminate range and is adequate for use in risk stratification. The functional sensitivity of the Bayer Centaur cTnI assay is at the low limit of the indeterminate zone, and values greater than 0.16 μg/L can be reliably used for such purposes.

Measuring cardiac troponins can facilitate early detection of ACS patients who are at high risk for adverse events. The key to translating such important findings into routine clinical practice is the ability to reliably measure cardiac troponin at concentrations near the lower limit of detection. Thus, the challenge for the diagnostic industry is to develop more sensitive cardiac troponin assays than those currently available; supersensitive assays can only improve precision in the low measuring range, resulting in reliable detection of high-risk ACS patients.

References

1. Katus HA, Scheffold T, Remppis A, Zehlein J. Proteins of the troponin complex. *Lab Med* 1992;23:311–317.
2. Dean KJ. Biochemistry and molecular biology of troponins I and T. In Wu AHB (ed): *Cardiac Markers*. Totowa, NJ: Humana Press; 1998:193–194.
3. Wu AHB, Valdes R Jr, Apple FS, et al. Cardiac troponin-T immunoassay for diagnosis of acute myocardial infarction and detection of minor myocardial injury. *Clin Chem* 1994;40:900–907.
4. Luscher MS, Thygesen K, Ravkilde J, Heickendorff L. Applicability of cardiac troponin T and I for early risk stratification in unstable coronary artery disease. The Thrombin Inhibition in Myocardial Ischemia Study Group. *Circulation* 1997;96:2578–2585.
5. Galvani M, Ottani F, Ferrini D, et al. Prognostic influence of elevated values of cardiac troponin I in patients with unstable angina. *Circulation* 1997;43: 2053–2059.
6. Wu AHB, Apple FS, Gibler WB, et al. National Academy of Clinical Biochemistry Standards of Laboratory Practice: Recommendations for use of cardiac markers in coronary artery diseases. *Clin Chem* 1999;45:1104–1121.
7. Fitzmaurice TF, Brown C, Rifai N, et al. False increase of cardiac troponin I with heterophilic antibodies. *Clin Chem* 1998;44:2212–2214.
8. Roberts WL, Calcote CB, De BK, et al. Prevention of analytical false-positive increases of cardiac troponin I on the Stratus II analyzer. *Clin Chem* 1997; 43:860–861.
9. Yeo KTJ, Storm CA, Li Y, et al. Performance of the enhanced Abbott AxSYM cardiac troponin I reagent in patients with heterophilic antibodies. *Clin Chim Acta* 2000;292:12–23.
10. Wu AHB, Feng YJ, Moore R, et al. Characterization of cardiac troponin subunit release into serum following acute myocardial infarction, and comparison of assays for troponin T and I. American Association for Clinical Chemistry Subcommittee on cTnI Standardization. *Clin Chem* 1998;44: 1198–1208.
11. Lindahl B, Venge P, Wallentin L. Relation between troponin T and the risk of subsequent cardiac events in unstable coronary artery disease. *Circulation* 1996;93:1651–1657.
12. Antman EM, Tanasijevic MJ, Thompson B, et al. Cardiac-specific troponin I levels to predict the risk of mortality in patients with acute coronary syndromes. *N Engl J Med* 1996;335:1342–1349.
13. Ohman EM, Armstrong PW, Christenson RH, et al. Risk stratification with admission cardiac troponin T levels in acute myocardial ischemia. *N Engl J Med* 1996;335:1333–1341.
14. Newby LK, Christenson RH, Ohman EM, et al. Value of serial troponin T

measures for early and late risk stratification in patients with acute coronary syndromes. *Circulation* 1998;98:1853–1859.
15. Wu AHB, Lane PL. Metaanalysis in clinical chemistry: Validation of cardiac troponin T as a marker for ischemic heart diseases. *Clin Chem* 1995;41: 1228–1233.
16. Olatidoye AG, Wu AHB, Feng YJ, Waters D. Prognostic role of troponin T versus troponin I in unstable angina pectoris for cardiac events with meta-analysis comparing published studies. *Am J Cardiol* 1998;81: 1405–1410.
17. Christenson RH, Duh SH. Evidence based approach to practice guides and decision thresholds for cardiac markers. *Scand J Clin Lab Invest* 1999; 59(suppl 230):90–102.
18. Lindahl B, Venge P, Wallentin L. Troponin T identifies patients with unstable coronary artery disease who benefit from long-term antithrombotic protection. *J Am Coll Cardiol* 1997;29:43–48.
19. Hamm CW, Heeschen C, Goldmann B, et al. Benefit of abciximab in patients with refractory unstable angina in relation to serum troponin T levels. *N Engl J Med* 1999;340:1623–1629.
20. Heeschen C, Hamm CW, Goldmann B, et al. Troponin concentrations for stratification of patients with acute coronary syndromes in relation to therapeutic efficacy of tirofiban. *Lancet* 1999;354:1757–1762.
21. Rodbard D. Statistical estimation of the minimal detectable concentration ("sensitivity") for radioligand assays. *Anal Biochem* 1978;90:1–12.
22. Hay ID, Bayer MF, Kaplan MM, et al. American Thyroid Association assessment of current free thyroid hormone and thyrotropin measurements and guidelines for future clinical assays. *Clin Chem* 1991;37:2002–2008.
23. Spencer CA, Takeuchi M, Kazarosyan M. Current status and performance goals for serum thyrotropin (TSH) assays. *Clin Chem* 1996;42:140–145.

Chapter 4

Report on a Survey of Analytical and Clinical Characteristics of Commercial Cardiac Troponin Assays

Fred S. Apple, PhD, Jesse E. Adams, III, MD, Alan H.B. Wu, PhD, and Allan S. Jaffe, MD

It has been shown that there is substantial variability in the analytical and clinical characteristics of cardiac troponin assays. However, there is a surprising lack of published information concerning these very substantial differences.[1,2] Clinical and analytical issues regarding biomarkers, especially cardiac troponin, were the topics of two annual meetings sponsored by the Jewish Hospital Heart and Lung Institute, in Louisville, Kentucky (October 1998 and 1999). The objective of this chapter is to compile the analytical and clinical characteristics of commercially available cardiac troponin I (cTnI) and T (cTnT) immunoassays. A survey was mailed to all manufacturers that have Food and Drug Administration (FDA)-approved cardiac troponin immunoassays to document analytical and clinical characteristics, including lower limit of detection, imprecision (coefficient of variation; CV%), upper reference limits, antibody configuration, acceptable specimen types, known interferences, and published clinical decision cutpoints. Two weeks after the surveys were mailed, each company was contacted by telephone for the purpose of requesting completion of the survey. Five out of 10 manufacturers responded. Five manufacturers' assays were also added to the database, based on package insert information. Thus, 10 manufacturers representing 16 different immunoassay systems are the basis of data presented in this report.

The analytical and clinical variability between the different assay systems is shown in Table 1. Time to first results ranged between 8

From: Adams JE III, Apple FS, Jaffe AS, Wu AHB (eds). *Markers in Cardiology: Current and Future Clinical Applications.* Armonk, NY: Futura Publishing Company, Inc.; © 2001.

Table 1

Analytical and Clinical Characteristics of Commercial Cardiac Troponin Assays

Assay		Ab Pair	Gen	Time to 1st Result (min)	Specimen Type	LLD	Ref Limit	Conc at 10% CV	ROC[a] Cutpoint
*Abbott:	AxSYM	M, G	1	8	S, HP	0.14	<0.50	N/A	2.0
Bayer:	Immuno 1	M, G	1	23	S, HP	0.10	<0.10	0.3	0.9
	ACS:180	M, G	1	15	S, HP	0.03	<0.10	0.2	1.0
	Centaur	M, G	1	15	S, HP	0.02	<0.10	0.2	1.0
*Beckman:	Access	M, M	1	15	S, EP	0.03	<0.03	N/A	0.15
*Biosite:	Triage	M, G	1	15	HP, WB	0.19	<0.19	2.0	0.4
Dade Behring:	Stratus II	M, M	1	10	S, HP	0.35	0.60	1.2	1.5
	Dimension RxL	M, M	2	16	S, HP	0.04	0.05	0.4	1.5
	Stratus CS	M, M	2	13	S, HP, WB	0.03	0.06	0.13	1.5
	Opus/Opus Plus	M, M	2	12	S, HP	0.10	<0.10	0.3	1.5
DPC:	Immulite	M, G	1	42	S, HP, EP	0.20	1.0	0.2	N/A
	Immulite Turbo	M, G	1	14	S, HP, EP	0.50	1.0	0.7	N/A
*First Medical:	Alpha Dx	M, M	1	18	S, HP, WB	0.09	<0.09	0.3	0.4
Ortho:	Vitros ECi	M, G	1	16	S, HP, EP	0.02	0.10	0.33	1.0
Roche:	Elecsys - 2010/1010	M, M	3	12	S, EP, CP	0.01	0.03	0.05	0.1
*Tosoh:	AIA	M, M	1	18	S, HP	0.45	<0.45	N/A	N/A

Roche assay is cTnT; all concentrations are expressed in ng/mL; *data obtained from package insert; S = serum; HP = heparin plasma; WB = whole blood; EP = EDTA plasma; CP = citrate plasma; M = Mouse monoclonal; G = goat polyclonal; N/A = not available; a = based on serum; Conc = concentration; Ref = reference; Gen = generation; Ab = antibody; LLD = lower limit of detection; EDTA = ectetic acid.

and 42 minutes. The lower limit of analytical detection ranged from 0.01 μg/L to 0.5 μg/L, a 50-fold difference. The lowest concentration giving a total imprecision (CV%) of 10% ranged from 0.05 μg/L to 1.2 μg/L, a 24-fold difference. Upper reference limits for a normal, healthy population defined at either the 97.5th or the 99th percentile ranged from 0.01 μg/L to 0.60 μg/L, a 60-fold difference. Not all assays had a documented upper reference limit. Receiver operating characteristic (ROC) curve cutpoints described for optimal sensitivity and specificity for the detection of acute myocardial infarction were available for only 13 assays. Three assays (Tosoh AIA [Tokyo, Japan], DPC Immulite, and DPC Turbo [Diagnostic Products Corporation, Los Angeles, CA]) had no clinical information available. ROC curve cutpoints ranged from 0.10 μg/L to 2.0 μg/L, a 20-fold difference.

Although several manufacturers reported data submissions pending approval by the FDA, only Dade Behring (Deerfield, IL), Roche Diagnostics (Indianapolis, IN), and Bayer Diagnostics (Tarrytown, NY), using cutpoints of 0.4 μg/L, 0.06 μg/L, and 0.1 μg/L, respectively, have FDA approval for use of their cardiac troponin assays for risk stratification in acute coronary syndrome patients.

The majority of assays were first generation. Dade Behring and Roche Diagnostics market second- and third-generation assays, respectively. In both cases, the later-generation assays have demonstrated improved analytical sensitivity, precision, and specificity compared with their first-generation systems. Three systems, Stratus CS (Dade Behring), Triage (Biosite Diagnostics Inc., San Diego, CA), and Alpha Dx (First Medical, Mountain View, CA), are capable of using anticoagulated whole blood as a specimen. All systems use either serum or plasma. Although this is not recorded by several manufacturers, it has been the users' experience that most assay systems show a lower (5% to 30%) bias for heparinized plasma compared with serum. It is recommended that if published information is not available from the assay manufacturer, each user validate any serum/plasma/whole blood bias if varied specimen types are to be used in clinical practice.

All assay systems used a mouse monoclonal antibody as the primary capture antibody. However, the secondary detection antibody varied between a mouse monoclonal antibody or goat polyclonal antibodies. As previously reported, the instability of circulating cTnI isoforms gives rise to both C- and N-terminal cTnI degradation.[3,4] Therefore, the user should request information from the manufacturer as to the epitope regions detected by each cTnI assay in order to recognize potential assay performance issues regarding in vivo and or in vitro cTnI instability. Recent reports have documented that several assays have interferences due to heterophile antibodies or rheumatoid factors[5,6] involving the Abbott (Abbott Park, IL) and Beckman (Fullerton,

CA) assays. Fibrin in serum has also been reported as a possible interferent for several systems. Serum pretreatment, serum filtration, or a hard spin before analysis has been shown to be successful in removing this potential interferent.[7]

This summation of surveyed data clearly documents the large clinical and analytical variability that exists between the many FDA-approved cardiac troponin assays in the worldwide marketplace. The lack of a repository for this sort of information constrains the field and causes confusion about how best to use these markers. The laboratory medicine and cardiology communities must understand the importance regarding the characteristics of the particular assay used in one's own laboratory for clinical practice. The interchanging of results between most systems remains almost impossible, and this likely leads to confusion regarding clinical interpretation. It is therefore critically important that each system be thoroughly evaluated both analytically and clinically if appropriate peer-reviewed literature is not available. The efforts of standardization and validation of cardiac troponin assays by subcommittees within the American Association for Clinical Chemistry and the International Federation of Clinical Chemistry will, we hope, partially or totally achieve these goals. We urge the cooperation of all manufacturers in this effort. The willingness of those who have shared their data is a good start.

References

1. Wu AHB, Apple FS, Gibler WB, et al. National Academy of Clinical Biochemistry standards of laboratory practice: Recommendations for use of cardiac markers in coronary artery diseases. *Clin Chem* 1999;45:206–212.
2. Apple FS. Clinical and analytical standardization issues confronting cardiac troponin I. *Clin Chem* 1999;45:18–20.
3. Katrukha AG, Berezikova AV, Filatov VL, et al. Degradation of cardiac troponin I: Implication for reliable immunodetection. *Clin Chem* 1998;44:2433–2440.
4. Morjana NA. Degradation of human cardiac troponin I after myocardial infarction. *Biotechnol Appl Biochem* 1998;28:105–111.
5. Dasgupta A, Banerjce SK, Datta P. False-positive troponin I in the MEIA due to the presence of rheumatoid factors in serum. *Am J Clin Pathol* 1999;112:753–756.
6. Fitzmaurice TF, Brown C, Rifai N, et al. False increase of cardiac troponin I with heterophile antibodies. *Clin Chem* 1998;44:2212–2214.
7. Roberts WL, Calcote CB, Holmstrom V, et al. Prevention of analytical false positive increases of cardiac troponin I on the Stratus II analyzer. *Clin Chem* 1997;43:860–861.

Chapter 5

The Current Assessment of Qualitative and Quantitative Point-of-Care Testing of Cardiac Markers

Roland Valdes, Jr., PhD and Saeed A. Jortani, PhD

Introduction

The use of biochemical markers for the diagnosis of acute coronary syndromes has become an integral part of current clinical practice. The rapid delivery of these tests is thought to reduce cost and improve the outcome of patients with chest pain who come to emergency rooms. Point-of-care testing (POCT) of myoglobin, cardiac troponin I (cTnI), cardiac troponin T (cTnT), and the MB isoenzyme of creatine kinase (CK-MB) has been developed and marketed with the promise of faster turnaround time (TAT). However, several issues, such as the usefulness of this mode of testing for the emergency department setting, the appropriate reporting format (qualitative versus quantitative), the responsible entity for monitoring the quality of such testing, the reporting and documentation of results, and the overall effects on patient outcomes, have not been fully resolved. In this chapter, we address some of these issues in the hope that their consideration will allow for a better understanding of the limitations and value of POCT for cardiac biomarkers.

Rapid advancement in the discovery and use of biomarkers for acute coronary syndromes has coincided with an intense effort for the development and implementation of POCT in general. POCT tests are diagnostic procedures performed at the bedside or near the patient.[1] Instead of samples being collected and delivered to the laboratory, the

From: Adams JE III, Apple FS, Jaffe AS, Wu AHB (eds). *Markers in Cardiology: Current and Future Clinical Applications.* Armonk, NY: Futura Publishing Company, Inc.; © 2001.

Table 1

Desired Attributes of an Ideal Point-of-Care Test

- Accurate
- Precise
- **Rapid (fast TAT)**
- Easy sample
- Low complexity
- Low calibration demands
- Disposable or low maintenance device
- **Low test cost**
- **Improved clinical decision making**

TAT = turnaround time.

testing is performed on site by a person involved in the care of the patient (such as a nurse or a physician). The attributes of an ideal POCT program are listed in Table 1. It is evident that faster assay TAT, lower test cost, and improved clinical decision making are the most desired attributes when considering POCT to replace testing by the traditional central laboratory. Whether these goals are currently being met by POCT for cardiac biomarkers remains unknown.

Devices made for POCT are usually small, portable, and simple to operate, and usually use whole blood as the specimen of choice. Depending on the modes of sample collection and analysis, POCT can be divided into in vitro, in vivo, and ex vivo types.[2] The in vitro technique involves removing a sample from the patient by using a test tube and then transferring the sample to the testing device. The measurement of analyte directly in the patient via intravenous, intra-atrial, subcutaneous, or transcutaneous routes is the in vivo testing type. The ex vivo testing type refers to situations in which the sample is withdrawn from the patient, and the analysis is done next to the patient (for example, an intra-atrial catheter that is attached to an analyzer).[1,2] Most cardiac biomarkers are currently in the in vitro format. The in vivo or the ex vivo types could potentially provide continuous monitoring of cardiac biomarkers in the blood. This continuous monitoring may allow for the better characterization of the rise and the fall in the levels of biomarkers in plasma, which may have potential clinical utility in the diagnosis of cardiac injury.

Cost containment is an important decision factor in the type of test chosen or the platform selected in the current delivery of cardiac biomarker testing. The notion that POCT costs less because it involves fewer steps and thus less labor seems logical. However, published studies on the economic aspects of POCT have generated an unclear picture of the subject, due to the inclusion or exclusion of many factors in these studies. Some studies have evaluated the cost of POCT by a single device and have compared it with the cost in the central laboratory.[3] In the "real-world" practice of POCT implementation, however, to achieve the promised faster TAT, many of the same types of POCT devices are placed throughout the facility.[4] The presence of many devices leads to the increased overall cost of maintenance, quality control, proficiency testing, and training. Unless all laboratory testing has been changed to the POCT, the cost of other tests still being performed in the central laboratory may increase if the high-volume cardiac biomarker tests are removed from the laboratory, and this would result in lower total annual volumes. This decrease in the number of biomarker tests done in the lab is not helpful in negotiating lower costs for all laboratory tests and could potentially offset the savings brought about by the switch to POCT. Therefore, it is prudent that the analysis of cost for cardiac markers be done by considering all components, including the time spent on the test by the medical staff, the overall effect on all laboratory testing, training, and quality control, and the use of multiple devices.

Some POCT devices generate a printout of test results that may be used for record keeping, whereas other devices do not and the result is usually read by subjective observation of the presence or absence of a line in the test area of the device. In busy emergency departments, where caregivers are responsible for many crucial tasks at any given moment, proper documentation and recording of test results onto the chart may inherently take lower priority. Furthermore, in such hectic environments, personnel may not comply properly with the quality control and with pre-analytical procedures intended to assure the accuracy of test results.[4] Therefore, in the course of evolution of cardiac biomarker testing from the central laboratory to the patient's bedside, efforts to automate record keeping (such as bar-coding) should also be given proper attention.

In this chapter, we address the following questions: 1) Is there a place for POCT for biomarkers in the diagnosis of coronary syndromes? 2) What is the current practice in POCT for cardiac markers? 3) Should results in POCT be qualitative or quantitative? 4) What tests should be offered in the POCT format? 5) Has the implementation of POCT for cardiac biomarkers changed patient outcomes?

Is There a Place for POCT for Biomarkers in the Diagnosis of Coronary Syndromes?

In cases in which electrocardiography (ECG) is diagnostic of acute myocardial infarction, patients are immediately treated before the results of biomarker testing are available.[5] The use of biochemical markers in these situations has a confirmatory role, and the peak (highest) value of biomarkers may provide a rough estimate of the infarct size.[6] In approximately 50% of the cases, ECG is nondiagnostic, and an objective alternative (that is, measuring biochemical markers in blood) is useful in aiding clinicians in their decision making about patient treatment and disposition.[5] Rapid TAT for the testing of biomarkers can reduce any delay in treatment of the patient and allow for triage and discharge decisions to be made more efficiently.[6] Therefore, it is apparent that one of the driving forces for the advancement of POCT has been the promise of a faster TAT (Table 1). It is recommended that results for STAT cardiac markers be available in 1 hour or less and that POCT should be implemented when this target cannot be consistently met.[6]

It is most appropriate to consider the therapeutic TAT when comparing the utility of POCT to that of traditional laboratory testing in triaging chest pain patients. According to Kost et al.,[1] therapeutic TAT is composed of the time needed 1) for ordering the test and getting the result to the clinical team, and 2) for initiation of medical or surgical treatment (Fig. 1). POCT potentially reduces the first component of therapeutic TAT, which is the time interval between ordering the test and getting the result to the clinician. POCT can achieve this faster TAT by having fewer required steps as compared with traditional laboratory testing.[7] Some of the time-consuming steps that are not part of POCT include: placing the order by the unit clerk, transporting the sample to the laboratory, processing the sample in the laboratory including centrifugation to obtain serum or plasma, pulling the order, reporting, and calling the physician.[7] It has been suggested that even though nurses usually perform the POCT, the actual time they spend on it is less than the time spent on sample collection, processing, transporting, or calling the physicians with the results.[7,8] The implementation of POCT for cardiac biomarkers also makes these tests available at more sites within a clinical facility. Faster TAT and availability at more sites brought about by POCT can potentially add more value to the clinical utility of cardiac biomarkers. The expected outcome is a more rapid initiation of therapy and ultimately an improvement in patient survival.

What is the Current Practice in POCT for Cardiac Markers?

We have summarized the data for proficiency testing of CK-MB and cTnI from a survey sponsored by the College of American Patholo-

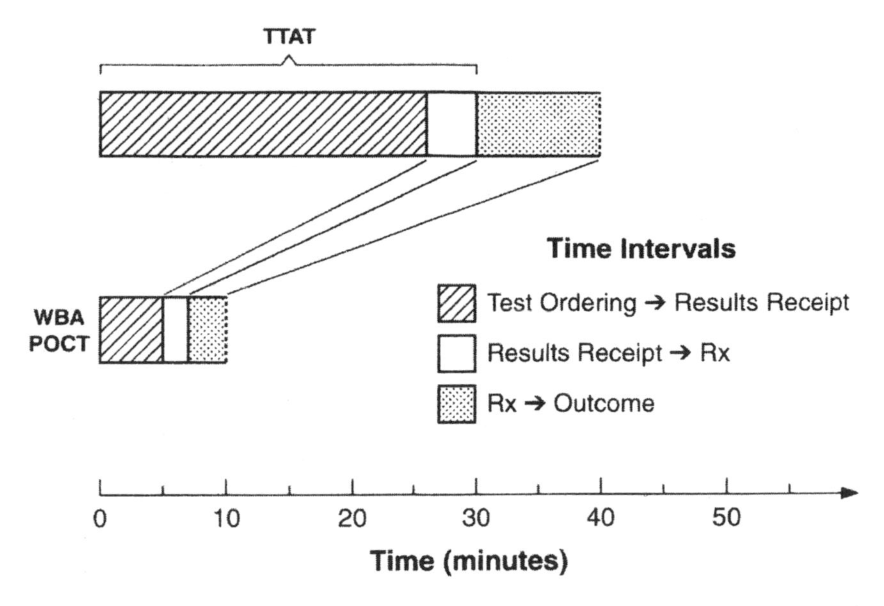

Figure 1. The concept of therapeutic turnaround time (TTAT). The time interval between the ordering of the test by the physician and the time at which results are available to the clinical team is the first component of TTAT. The other component is the time needed by the clinical team to initiate therapy, either medical or surgical intervention. The time for the first component of TTAT can potentially be shortened by point-of-care testing (POCT). WBA = whole blood assay. From Kost G et al. *Chest* 1999;115:1140–1154.

gists in 1998 and in 1999 (Fig. 2). In this figure, we present the data for a set of samples sent in March of each year.[9,10] It is important to note that numbers represent the laboratories reporting. Since laboratories have variable test volumes, the actual number of tests performed for each marker may vary. However, the data presented provide a rough assessment of the current practice in cardiac biomarker testing. The number of laboratories performing CK-MB testing in 1999 increased modestly (by 18%), whereas laboratories reporting cTnI increased by 2.5-fold (Fig. 2). In the March 1998 survey, there were no laboratories that reported POCT qualitative CK-MB or cTnI results. In the March 1999 survey, there were 23 respondents using laboratory-based testing and 12 respondents using POCT devices who reported their CK-MB results qualitatively. It is also apparent that there were more laboratories performing cardiac biomarker testing in 1999 than in the previous year.

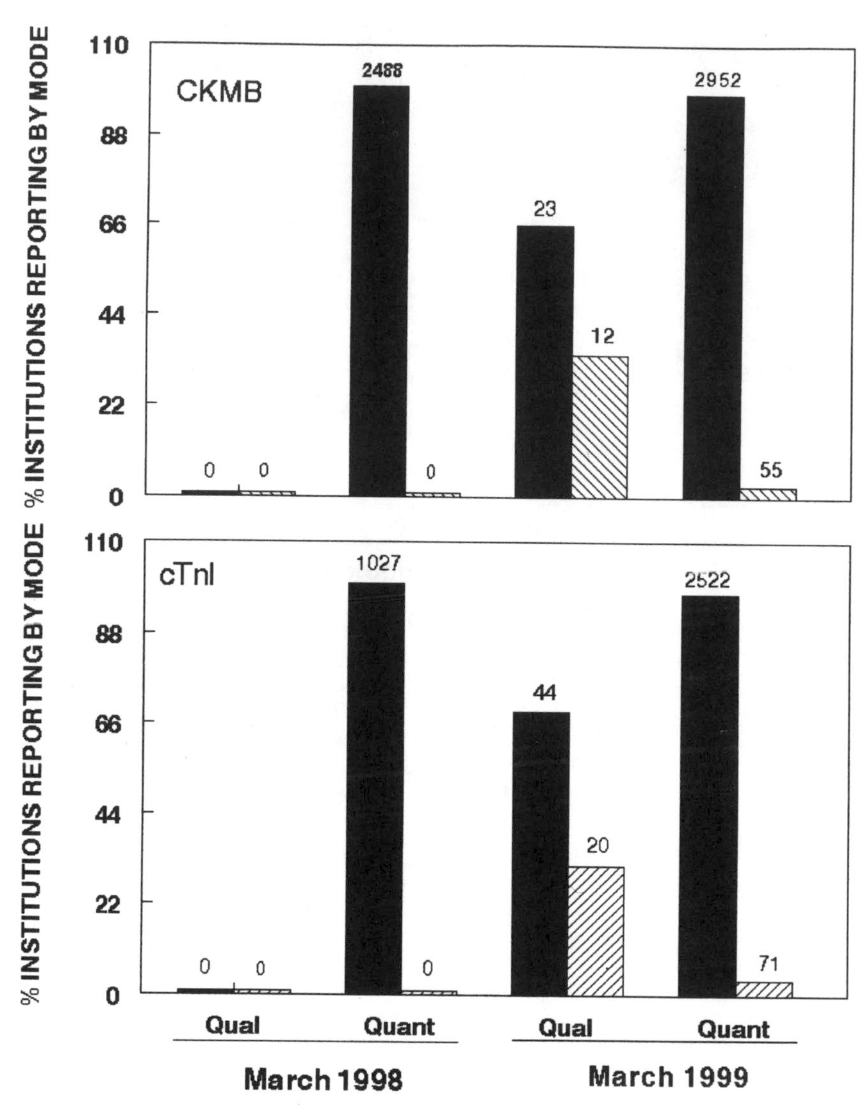

Figure 2. College of American Pathologists proficiency testing surveys for creatine kinase MB (CK-MB; top) and cardiac troponin I (cTnI; bottom) in March 1998 and March 1999. The point-of-care tests are represented by bars with crossed lines. The tests performed in the laboratory are represented as solid black bars. The percent of respondents reporting by mode is shown on the y-axis. Note that there were no laboratories reporting qualitative results for either CK-MB or myoglobin in the 1998 survey. Also note that the majority of qualitative testing is still performed in the central laboratory. The total actual number of respondents is shown above the bars.

Should Results in POCT be Qualitative or Quantitative?

The issue of reporting mode (that is, quantitative versus qualitative) should be addressed before the development, implementation, and utilization of POCT for cardiac biomarkers. Various factors such as availability, clinical need for quantitative results, and cost may influence the decision making relative to the type of POCT chosen. Table 2 lists cardiac biomarkers that are currently available in POCT format in the US. Although qualitative POCT is offered for cTnT (Roche Diagnostics, Indianapolis, IN), and for CK-MB, myoglobin, and troponin I by the cardiac STATus devices (Spectral Diagnostics, Toronto, Canada), the majority of POCT for biomarkers currently offered are quantitative. The major issue to consider here is which mode should be used for the optimum reporting of results. Reporting the results of cardiac biomarker testing in quantitative terms seems to be more acceptable because when serial sampling is done, quantitative data are more informative. In addition, having the quantitative data available, the clinician gets a perspective on the extent of the injury based on the values for the tested biomarkers.[11–13] Thus, reporting in qualitative terms may limit the value of biomarker testing in clinical practice. We propose that the issue of analytical differences in generating quantitative or qualitative results should be considered when a reporting mode is chosen. Qualitative tests use a cut-off value, and reporting is done in an "all or none" fashion. A low correlation has been shown between POCT qualitative test and lab-generated quantitative values for cTnT for re-

Table 2

Current Methods for Point-of-Care Monitoring of Biomarkers in AMI (circa 1999)[a]

	Hybr Icon[b]	Card T Rpd	STATus	STATus TnI	TRIAGE	Alpha DX	RAMP	Stratus CS
CK						•		
CK-MB	•		X		•	•	•	•
Myo			X		•	•	•	•
cTnI				X	•	•	•	•
cTnT		X						

[a]Quantitative methods are shown as (•) and qualitative methods as (X). [b]The Hybritech Icon is serum-based. All others make use of whole blood. CK = creatine kinase; CK-MB = creatine kinase MB; Myo = myoglobin; cTnI = cardiac troponin I; cTnT = cardiac troponin T.

sults near the cut-off point.[14] Results close to the cut-off point, however, may be informative and should not be ignored.

What Tests Should be Offered in the POCT Format?

Recent advances in the development of new tests for biomarkers for cardiac injury have offered more options to clinicians for ruling out acute coronary syndromes. Use of the newer tests has also become possible at or near the patient's bed through the adaptation of the tests to POCT format. It is important to choose the most appropriate marker(s) to be included in the POCT format based on their inherent times of rise and fall after cardiac injury. Therefore, decisions concerning which assays are to be included in POCT must be made with the physiological and analytical limitations of each marker in mind. The National Academy of Clinical Biochemistry recommends that two biochemical markers should be used for the routine diagnosis of acute myocardial infarction.[6] Their recommendation is to have a marker that reliably increases in blood within 6 hours after the onset of symptoms (early marker) and a marker that is increased in 6 to 9 hours with high sensitivity and specificity for cardiac injury (definitive marker). It is therefore possible that an early marker such as myoglobin and a definitive marker such as cTnI or cTnT may be sufficient for POCT format. The role of CK-MB in POCT has now become questionable, and its future use may partially depend on the analytical and clinical success of troponins in POCT format.

Has the Implementation of POCT for Cardiac Biomarkers Changed Patient Outcome?

Outcome studies have attempted to address the analytical, diagnostic, and therapeutic aspects of POCT. Kendall et al.[15] designed an open, single-site, randomized, controlled trial with 1728 patients enrolled. Their objective for this study was to determine the extent of change resulting from the use of POCT in the management of patients in an emergency department setting. In this trial, they evaluated the use of i-STAT (Princeton, NJ) for the determination of Na^+, K^+, Cl^-, urea, glucose, packed cell volume, and hemoglobin concentrations. They also evaluated a second cartridge for measuring blood gases on the same instrument. Their main measures of outcome were mortality, length of hospital stay, admission rate, time spent waiting for laboratory results, time spent for decision making and management planning, and time spent in the emergency department. All parameters were compared for two groups of patients randomly assigned to POCT or to testing by the hospital's central laboratory. Results of their study indicated that timing was considered critical in the management of 859 patients out of the total number enrolled. Of these 859 patients, a critical change occurred in only 59 patients (6.9%) in the POCT group. Decisions were made 74 minutes earlier in the POCT group for hematology testing, 86 minutes earlier for chemistry testing, and 21 minutes earlier for blood gases than in the laboratory. Despite the increased efficiency of POCT, these time changes did not alter the clinical outcome or time spent in the emergency department. Parvin et al.[16] assessed the use of a hand-held POCT device for measuring electrolytes in an emergency department in a clinical trial involving 4985 patients. They concluded that the routine use of the POCT device had not affected the length of stay for patients in their large emergency department. Halpern et al.,[17] however, assessed the clinical and economic impact of a POCT device for measuring PO_2, pH, PCO_2, Na^+, K^+, ionized Ca^{2+}, and hematocrit in blood in 1590 patients after coronary artery bypass graft surgery. They found that the use of POCT correlated with decreased ventricular arrhythmias and postcardiac arrest events. Improvements in clinical outcomes were made that were accompanied by economic savings. These savings were due in part to the smaller amount of time spent in the intensive care unit by patients in the POCT group versus patients who had their testing performed in the STAT lab.

Dadkhah et al.[18] recently compared the use of qualitative POCT for myoglobin, CK-MB, and cTnI to the lab-based quantitative testing for the same analytes in decision making for patients with acute myocardial infarction. They enrolled 127 consecutive patients with nondiagnostic ECG, of whom 118 had negative serial cardiac markers and 9 had positive results. Six of the nine had been treated based on the POCT result, and the diagnostic correlation between the lab-based quantitative values and the POCT-generated qualitative results was 100%. Therefore, they concluded that qualitative POCT for cardiac biomarkers was comparable to the laboratory-generated quantitative approach with the added advantage of faster TAT. Unfortunately, the number of patients in that study was small. Nevertheless, Brogan and Bock[19] have indicated that in situations where STAT quantitative testing for cardiac markers is available, the impact of POCT for these analytes would be expected to be less.

Heeschen and colleagues[20] have also evaluated the use of a POCT-based qualitative testing for cardiac biomarkers for the measurement of the same analytes (also quantitatively) in the laboratory. They enrolled 637 patients of whom 159 had myocardial infarctions. They determined a diagnostic correlation of 99% between the POCT and laboratory testing for the markers. They also concluded that POCT was comparable in its diagnostic capability to the results from the laboratory. In assessing the utility of the rapid cTnT assay in the emergency department, the Rapid Evaluation by Assay of Cardiac Troponin T (REACTT) group[21] concluded that the rapid whole-blood assay was comparable in sensitivity to the lab-based serum assay with the added advantage of a faster TAT. However, they noted discrepancies in the POCT-generated result versus the lab-based serum assay for cTnT that they attributed to the need for further education of the users of the test reader. In a different trial of POCT for cTnT, Muller-Bardorff et al.[22] assessed the use of a charge coupled device camera to measure the reflectance of the signal lines—an automated quantitation of results. They compared the quantitative results from the cTnT rapid assay using the reader with the quantitatively measured cTnT in the central laboratory. A total of 252 patients with acute coronary syndromes were enrolled to determine test efficiency, and 140 samples were tested for assay correlation. They found that analytically, the two assays correlated ($y = 0.85X + 0.002$ μg/L; $r = 0.98$), and the POCT quantitative result was more rapid. Diagnostic performance of qualitative POCT testing for cTnT has been shown to be adequate or slightly lower than lab-based quantitative testing for this particular analyte in several other studies.[23,24]

Although the great advancements in the development of cardiac biomarkers and the recent proliferation of POCT have raised everyone's expectations, it has not been substantiated whether such an approach has improved the ability to diagnose acute coronary syndromes to the extent that it would alter patient outcome. Many questions still remain and should be addressed before achieving the goal of rapid, accurate, and cost-effective monitoring of cardiac markers near the patient's bed. Improvement in the use of resources and in clinical outcome may be achieved through linking POCT with improvements in the triaging of patients and treatment strategies.[25] In any event, the clinical laboratory's roles in assuring the quality of POCT and in providing technical expertise to users will still be pivotal in the success of near-patient testing for cardiac biomarkers.

References

1. Kost GJ, Ehrmeyer SS, Chernow B, et al. The laboratory-clinical interface: Point-of-care testing. *Chest* 1999;115:1140–1154.
2. Kost GH, Hague C. In vitro, ex vivo, and in vivo biosensor systems. In Kost GH (ed): *Handbook of Clinical Automation, Robotics, and Optimization.* New York: John Wiley and Sons; 1996:648–753.
3. Nosanchuck JS, Keenfer R. Cost analysis of point-of-care laboratory testing in a community hospital. *Am J Clin Pathol* 1995;103:240–243.

4. Lyon AW. The luxury of bedside testing. *Am J Clin Pathol* 1995;104:107–108.
5. Hamm CW, Goldman BU, Heeschen C, et al. Emergency room triage of patients with acute chest pain by means of rapid testing for cardiac troponin T or troponin I. *N Engl J Med* 1997;337:1648–1653.
6. Wu AHB, Apple FS, Gibler WB, et al. Recommendations for the use of cardiac markers in coronary artery disease. In Wu AHB, Apple FS, Warshaw MM (eds): *Standards of Laboratory Practice*. Vol. 5. Franklin: Durik Advertising, Inc.; 1999:1–44.
7. Bailey TM, Tophram TM, Wantz S, et al. Laboratory process improvement through point of care testing. *Jt Comm J Qual Improv* 1997;23:362–380.
8. Harvey MA. Point-of-care laboratory testing in critical care. *Am J Crit Care* 1999;8:72–83.
9. College of American Pathologists, Proficiency Testing Surveys, Cardiac Markers Survey Set CAR-A (1998), Northfield, IL.
10. College of American Pathologists, Proficiency Testing Surveys, CAR-A set (1999), Northfield, IL.
11. Omura T, Teragaki M, Takagi M, et al. Myocardial infarct size by serum troponin T and myosin light chain 1 concentration. *Jpn Circ J* 1995;59: 154–159.
12. Delanghe JR, De Mol AM, De Buyzere ML, et al. Mass concentration and activity concentration of creatine kinase isoenzyme MB compared in serum after acute myocardial infarction. *Clin Chem* 1990;36:149–153.
13. Voss EM, Sharkey SW, Gernert AE, et al. Human and canine cardiac troponin T and creatine kinase-MB distribution in normal and diseased myocardium. Infarct sizing using serum profiles. *Arch Pathol Lab Med* 1995;119: 799–806.
14. Luscher MS, Ravkilde J, Thygesen K. Clinical application of two novel rapid bedside tests for the detection of cardiac troponin T and creatine kinase-MB/myoglobin in whole blood in acute myocardial infarction. *Cardiology* 1998;89:222–228.
15. Kendall J, Reeves B, Clancy M. Point of care testing: Randomized controlled trial of clinical outcome. *Br Med J* 1998;316:1052–1057.
16. Parvin CA, Lo SF, Deuser SA, et al. Impact of point-of-care testing on patients' length of stay in a large emergency department. *Clin Chem* 1996; 42:711–717.
17. Halpern MT, Palmer CS, Simpson KN, et al. The economic and clinical efficiency of point-of-care testing for critically ill patients: A decision-analysis model. *Am J Med Qual* 1998;13:3–12.
18. Dadkhah S, Fisch C, Zonia C, Foschi A. Accelerated coronary reperfusion through the use of rapid bedside cardiac markers. *Angiology* 1999;50:55–62.
19. Brogan GX, Bock JL. Cardiac marker point-of-care testing in the emergency department and cardiac care unit. *Clin Chem* 1998;44:1865–1869.
20. Heeschen C, Goldman BU, Moller RH, Hamm CW. Analytical performance and clinical application of a new rapid bedside assay for the detection of serum cardiac troponin I. *Clin Chem* 1998;44:1925–1930.
21. Baxter MS, Brogan GX Jr, Harchelroad FP, et al , for the REACTT Investigators Study Group. Evaluation of a bedside whole-blood rapid troponin T assay in the emergency department. *Acad Emerg Med* 1997;4:1018–1024.
22. Muller-Bardorff M, Rauscher T, Kampmann M, et al. Quantitative bedside assay for cardiac troponin T: A complementary method to centralized laboratory testing. *Clin Chem* 1999;45:1002–1008.
23. Gerhardt W, Ljungdahl L, Collinson PO, et al. An improved rapid troponin

T test with a decreased detection limit: A multicenter study of the analytical and clinical performance in suspected myocardial damage. *Scand J Clin Lab Invest* 1997;57:549–558.
24. Sylven C, Lindahl S, Hellkvist K, et al. Excellent reliability of nurse-based bedside diagnosis of acute myocardial infarction by rapid dry-strip creatine kinase MB, myoglobin, and troponin T. *Am Heart J* 1998;135:677–683.
25. Hudson MP, Christenson RH, Newby LK. Cardiac markers: Point of care testing. *Clin Chim Acta* 1999;284:223–237.

Chapter 6

Acute Coronary Syndromes: *Pathophysiology, Clinical Presentation, and Initial Diagnostic Strategies*

Jesse E. Adams, III, MD and Vickie A. Miracle, RN, EdD

Introduction

To properly understand the application of currently available and future markers of myocardial injury and necrosis, one must possess an understanding of the scope of the problem as well as the opportunities and limitations of the current management model. Coronary artery disease is the number one cause of mortality in the US, where one person dies from cardiac disease every 34 seconds.[1] Up to 8 million Americans are seen in emergency departments (EDs) each year complaining of chest discomfort, and 2 to 3 million are admitted to a hospital having been diagnosed with acute coronary syndrome (ACS). More than 1,500,000 people suffer an acute myocardial infarction (AMI) each year (many without seeking medical care), and one third of these people die. It is estimated that up to 4% (most commonly reported, 1% to 2%) of individuals with AMIs are discharged inappropriately from the ED. Reported mortality rates in patients who are inappropriately discharged appear to be higher than for those patients who are admitted.[2] The public and health care providers must promptly identify symptoms of ACS to initiate appropriate treatment because myocardial ischemia and cell death often progress rapidly. Therefore, therapeutic methods have the greatest impact in the first few hours after the onset of ACS. An appropriate diagnostic strategy must have a high negative

From: Adams JE III, Apple FS, Jaffe AS, Wu AHB (eds). *Markers in Cardiology: Current and Future Clinical Applications.* Armonk, NY: Futura Publishing Company, Inc.; © 2001.

predictive value to maximize the detection of patients with ACS in a cost-efficient manner. The achievement of this goal is complicated by the fact that fewer than 25% of individuals admitted to the hospital with chest discomfort are ultimately found to have myocardial necrosis.[3]

The inappropriate discharge of patients with ACS has often been avoided by using a very low threshold for hospital admission. Historically, up to 70% or more of patients who present to the ED with chest discomfort will be admitted for further evaluation ("rule out MI"). Often only 15% to 30% of these patients will have suffered any myocardial injury, and usually one third to one half of these individuals will be determined to have a noncardiac cause of their chest pain. Cost constraints in health care have created an impetus to treat patients with chest pain in a cost-efficient manner. To contain costs without sacrificing quality, it is necessary to have a superior understanding of the spectrum of ACS along with a protocol-driven approach to assess individual patients in a systematic way. Opportunities thus exist for improved public understanding to facilitate early presentation as well as for improved diagnostic tools to enable health care providers to detect myocardial ischemia and infarction more rapidly and with greater reliability.

Pathophysiology

ACS is a continuum of a pathologic process originating in the epicardial coronary arteries but involving an entire circulatory distribution, resulting in cellular ischemia in the myocardium distal to the vascular lesion. The clinical presentation that subsequently results can be quite variable. Silent ischemia and infarction are common, but many patients will have no symptoms at all. It should be stressed that silent ischemia is not limited to diabetics; episodes of silent ischemia are common in all patients with coronary artery disease. Separate "points" on this spectrum have been labeled as unstable angina, non-Q-wave MI, and Q-wave MI, but it must be recognized that all have a common pathologic process that is quite complex.[4] The size of the thrombus formed and the resulting degree of blood supply impairment largely determine the diverse clinical diagnoses (unstable angina, non-Q-wave MI, Q-wave MI) (Fig. 1). The severity of ischemia will be defined by the degree of luminal narrowing of the epicardial artery, the presence or absence of collateral formation, the degree of ventricular hypertrophy, other determinants of myocardial oxygen demand (such as heart rate, thyroid status, level of circulating catecholamines, etc.), and oxygen supply (such as degree of oxygenation, oxygen carrying capacity, etc.).

The triggering events and endothelial pathology, however, are

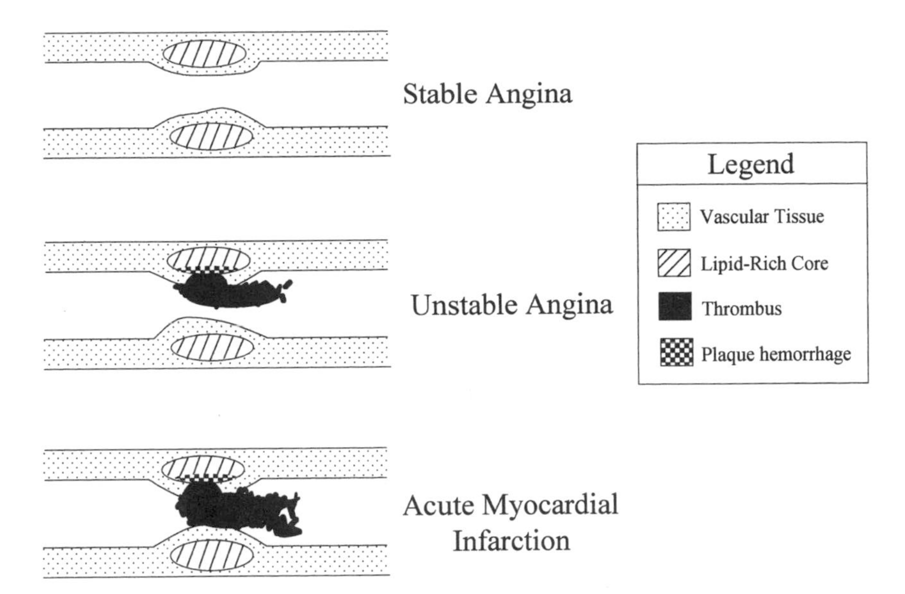

Figure 1. The clinical presentations of stable angina, unstable angina, and acute myocardial infarction are ultimately determined by disease involving the endothelial surface of an epicardial artery.

similar. It should be noted that patients with ACS have varying degrees of "ischemia," sometimes with conjoint cellular necrosis. The term ischemia must be used carefully and with a correct understanding of its context. We can correctly refer to individuals with stable angina pectoris or positive exercise stress tests as having episodes of myocardial ischemia. We can also correctly designate individuals who present with unstable angina pectoris or MI as having episodes of myocardial ischemia. However, ischemia in the setting of stable angina pectoris and ischemia after the onset of unstable angina are different pathologic entities, and discussions involving tests for "ischemia" should not confuse the two. In patients with stable angina or in individuals who have had a positive stress evaluation, there is typically no ruptured atherosclerotic plaque, nor is there concurrent activation of thrombotic and cellular cascades. Conversely, individuals who present with ACS or AMI typically have plaque rupture with a significant vascular endothelium perturbation in addition to the cellular ischemia.

ACS is caused by an imbalance in blood flow/oxygen supply and oxygen demand in part of the myocardium, often due to atherosclerosis. The biological details of atherosclerotic plaque formation and rupture are exceedingly complex. For a more complete explanation, please refer to a recent American Heart Association monograph on the subject.[5] While five separate causes of unstable angina have been recently enumerated, the event is most commonly initiated by the rupture of an unstable lipid-filled plaque in an epicardial coronary artery.[6] The plaque is typically an advanced atherosclerotic lesion covered by a fibrous cap consisting of smooth muscle cells and connective tissue containing macrophages and T lymphocytes (Fig. 2). Under the fibrous cap, the lipid-rich core is filled with macrophages, lipids, calcium, and other materials. Unstable plaques often have activated macrophages and leukocytes on the "shoulders" of the plaque.

Certain characteristics of plaque may predispose rupture. These characteristics include a softer lipid core, the presence of macrophages, less smooth muscle cells, a thinner fibrous cap, and increased inflammatory activity. Although the precise mechanisms resulting in plaque rupture are still incompletely defined, increases in blood pressure, shear stress, capillary hemorrhage into the plaque, spasm, exposure to cold, emotional or physical stress, and hypercoagulable states have all been implicated.[7,8] Patients who have preinfarction unstable angina pectoris have been found to have evidence for a greatly enhanced acute phase (inflammatory) response with the elevation of inflammatory markers such as C-reactive protein, serum amyloid A protein, and interleukin 6.[9] Enhanced activation of the coagulation cascade has also been described in patients who have unstable angina and detectable levels

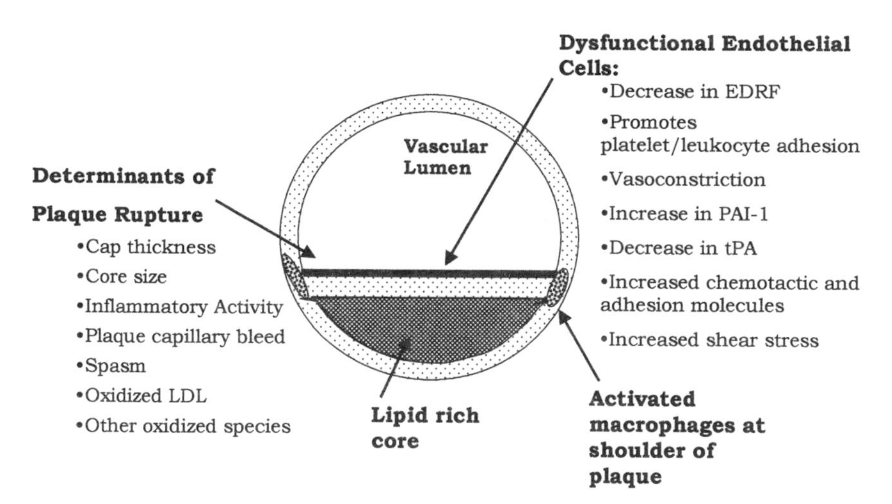

Figure 2. An atherosclerotic lesion is a complex involving a lipid-rich core, inflammatory cells, and alterations in the endothelial cell surface. Marked abnormalities in the production of circulating mediators are found. EDRF = endothelium-derived relaxing factor; LDL = low-density lipoprotein; PAI-1 = plasminogen activator inhibitor; tPA = tissue plasminogen activator.

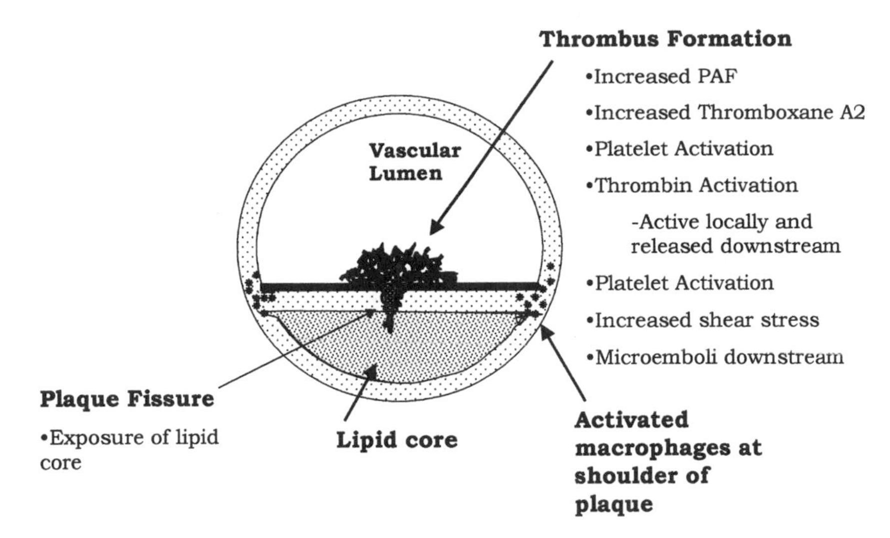

Figure 3. After plaque rupture occurs, a platelet-rich thrombus forms rapidly, with potential effects on lumen diameter, subsequent downstream perfusion, and potent procoagulant effects. Mediators, cellular ingredients, and microemboli are released with deleterious downstream effects. PAF = platelet activating factor.

of troponin.[10] Immediately after plaque rupture occurs, platelets are exposed to the lipid-rich core that is highly thrombogenic (Fig. 3). The platelets release thrombin, platelet activating factor, adenosine diphosphate, serotonin, and thromboxane A_2, all of which further stimulate platelet activation and recruitment. As glycoprotein IIb/IIIa receptor sites become activated, even more platelet activation occurs, facilitating the binding of fibrinogen/fibrin to platelets, vascular endothelium, media, and the components of the lipid-rich core.[8,10] Concurrently, the stimulation of leukocytes occurs, with the release of mediators such as tumor necrosis factor-α, macrocyte stimulatory factor, neutrophil chemotactic factors, and interleukin 8. Serum-derived factors such as complement split products (especially C3a and C5a) further stimulate leukocyte activity. These factors will all "turn up the heat" at the site of injury, rapidly escalating the propensity for clot formation, vasoconstriction, and downstream ischemia.

This procoagulant stimulation occurs frequently in the setting of diminished capacity for compensatory mechanisms. Endogenous fibrinolytic activity is diminished in atherosclerotic arteries. Both platelets and endothelial cells are significant producers of plasminogen activator inhibitor (PAI-1), the primary inhibitor of tissue plasminogen activator (tPA). An increased ratio of PAI-1 to tPA has been found to increase the risk of ACS. A circadian rhythm of PAI-1 levels has been observed, with peaks in the early morning hours. Higher levels have been described in diabetics as well. Dysfunctional endothelial cells overlying the fibrous cap also produce less endothelial-derived relaxation factor, which in turn promotes platelet and leukocyte deposition, increases shear stress, and facilitates vasoconstriction.

Once plaque rupture occurs and thrombus formation is under way, then the processes described briefly above will begin to affect blood flow past the site of plaque disruption. The size of the thrombus formed, the degree of distal embolization, the morphology and size of the residual lumen, the diameter of the stenosis, the presence and degree of additional coronary artery disease in other circulatory territories, and the presence or absence of collateral formation will all help to determine the degree of ischemia and necrosis. The occlusion is often intermittent due to competition between the thrombotic and fibrinolytic cascades, to the reversible nature of spasm, and to variations in myocardial oxygen demand. This intermittent occlusion is often manifested clinically in variations in the presence or severity of anginal chest pain, intermittent abnormalities seen on serial electrocardiograms (ECGs) or with continuous ST-segment monitoring, or in transient repeated elevations of small molecular weight molecules (ie, the "staccato phenomenon" observed with myoglobin).

Clinical Presentation

Once myocardial ischemia has begun, patients with ACS will often present with some complaint of chest discomfort. Classic symptoms include sternal or retrosternal chest discomfort which may radiate to the left arm, shoulder, neck, or back accompanied by dyspnea, diaphoresis, nausea, and a feeling of impending doom. While patients are usually asked if they have chest "pain," the discomfort is usually not truly painful. Instead, the discomfort is often described as pressure, heaviness, burning, tightening, stabbing, choking, crushing, or indigestion. However, not all patients will present with classic symptoms. Up to 30% of patients will have no discomfort at all and may present with an anginal equivalent: congestive heart failure, dyspnea, arrhythmias, lightheadedness, gastrointestinal distress, and/or dizziness. Women often present with less expected signs and symptoms and may complain of vague chest or upper back discomfort, fatigue, dyspnea, palpitations, dizziness, diminished exercise tolerance, dyspnea on performing routine chores (such as sweeping or vacuuming), or other atypical signs of coronary artery disease. The elderly also frequently present with such "atypical" symptoms. Thus, all complaints of chest discomfort must be taken seriously, but the absence of symptoms, even including "atypical" symptoms, does not allow the physician to exclude the presence of episodes of myocardial ischemia.[7]

In unstable angina, short episodes of chest discomfort occur (typically 30 seconds to 10 minutes) due to intermittent episodes of ischemia. The discomfort may be relieved with sublingual nitroglycerin or rest. Activity, emotional stress, cold weather, or any trigger that will increase myocardial oxygen demand may initiate the discomfort, but the discomfort may also occur at rest. Discomfort associated with AMI is more persistent and is not usually relieved with nitroglycerin or rest, or both. Despite popular public opinion, the majority of AMIs occur at rest, while sleeping, or with normal activity. In most population studies, two periods of increased incidence have been described, the largest in the early morning hours (often superimposed on the rise in catecholamines that occur at this time), and then a second peak in the late afternoon. Some have suggested that this pattern may not apply to all populations, and that there may be a difference with race. Some have found that those of African ancestry have the greatest incidence of MIs later in the day.

An evaluation of a patient's history and clinical presentation must also include an assessment of a patient's risk profile for the development of atherosclerosis. A number of risk factors that increase the likelihood for coronary artery disease have been described. These include: 1) hyperlipidemia defined as total cholesterol level greater than 200 mg/dL or low-density lipoprotein greater than 130 mg/dL or high-density lipoprotein less than 35 mg/dL; 2) tobacco smoking; 3) hypertension; 4) sedentary lifestyle; 5) heredity, particularly if a family member was diagnosed with coronary artery disease before the age of 60; 6) obesity; 7) diabetes mellitus; 8) stress; 9) high triglyceride levels; and 10) elevated homocysteine levels. As the number of risk factors present in an individual increases so does the likelihood of coronary artery disease. However, these risk factors often are poorly predictive for a cardiac cause of chest pain.

Ultimately, in most individuals, it is usually impossible to definitively determine if they have had episodes of myocardial ischemia that have resulted in symptoms or in chest discomfort. Most individuals who present to their physicians' office or to an ED with complaints potentially attributable to unstable angina will not have data obtained on their history that will conclusively answer the clinical question. Accordingly, further diagnostic testing is crucial.

Diagnostic Testing

To ascertain if an individual patient is suffering from an ACS, a stepwise progression of diagnostic tests is used (Fig. 4). The use of

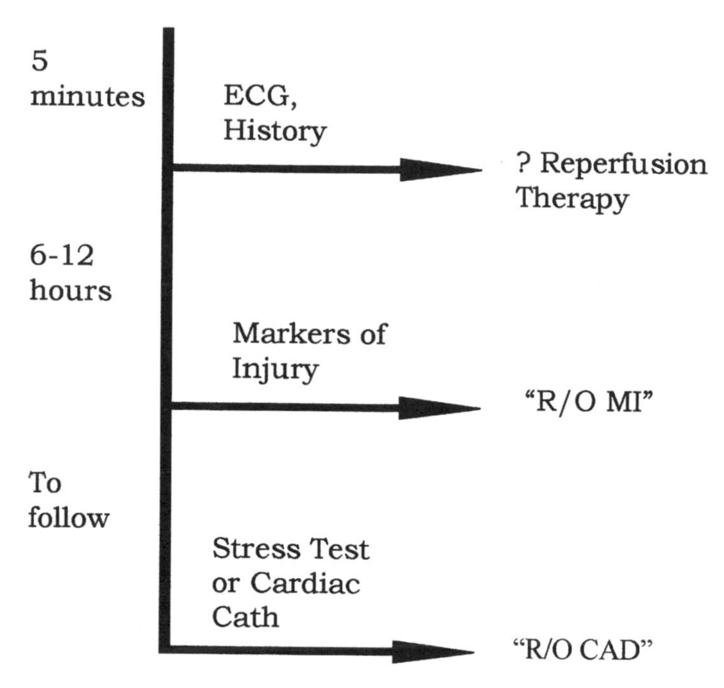

Figure 4. A protocol-driven approach is necessary for the optimum evaluation and treatment of patients who present to the emergency department with chest pain. Although a number of protocols have been developed, the basic needs consist of a three-step process as defined here. ECG = electrocardiogram; R/O MI = rule out myocardial infarction; R/O CAD = rule out coronary artery disease.

diagnostic tests in the ED should follow a logical progression that provides information to guide further evaluation and treatment. The rate of irreversible cellular injury progresses rapidly in patients with MI; most of the damage has occurred within the first 6 hours after coronary artery occlusion. However, with newer medications and procedures, the incidence of mortality and morbidity associated with ACS has decreased.[11] Thus, a prompt assessment for the presence of an acute transmural MI that would require reperfusion therapy (thrombolysis or cardiac catheter-based intervention) is indicated.[12]

Within 5 to 10 minutes of arrival in the ED, the patient with chest discomfort should have a directed history and physical obtained and an ECG performed and reviewed by the physician in the ED. If no diagnostic abnormalities are present, then it is necessary to evaluate for lesser degrees of myocardial necrosis that will have important implications on triage, treatment, and prognosis.

The detection of smaller degrees of myocardial necrosis is typically performed by the serial measurement of markers of cardiac injury (especially creatine kinase MB [CK-MB] or troponin). This is often referred to as a "rule-out protocol for MI." This rule-out protocol can be performed by a number of molecular markers of myocardial injury as discussed in this monograph. Currently, the measurement of troponins or CK-MB, or both, is performed; measurement of "early" markers such as myoglobin or CK-MB isoforms may be added to attempt to shorten the time to diagnosis (Fig. 5).[13,14] It is important that serial measurements be performed for any marker used.[15,16] It must be

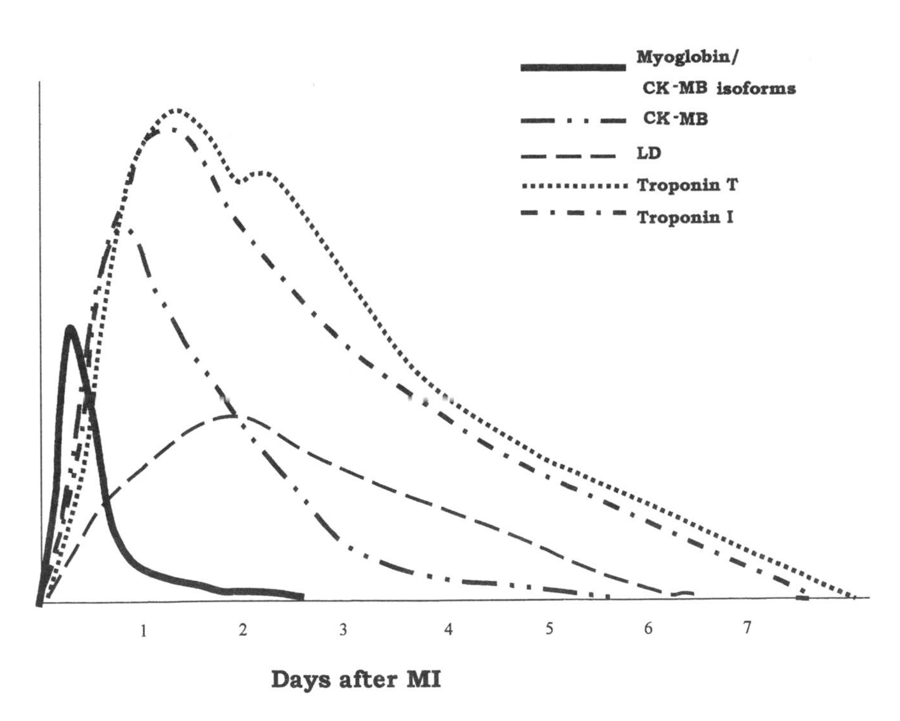

Figure 5. Serum markers follow a "rising and falling" pattern after myocardial cellular injury. In part, the duration of elevation will be affected by the size of the infarction and the presence or absence of reperfusion in the affected territory. MI = myocardial infarction; CK-MB = creatine kinase MB; LD = lactate dehydrogenase.

stressed that serial testing heightens the sensitivity of detection for any blood-based marker. Unfortunately, serial testing is still not the norm. Some studies have found that less than half of those patients who present with chest discomfort to the ED have serial measurements of CK-MB or troponins. For those hospitals with chest pain units, the numbers are still low, with only three quarters documenting serial testing. Additionally, the sensitivity for detection of myocardial necrosis is predicated on the time from the onset of chest pain to the time of presentation. If sufficient time has not elapsed to allow for the release of the marker into the bloodstream, then the level of the marker in the circulation may be normal despite the presence of myocardial necrosis. Additionally, the "clock" must be reset for any subsequent episode of chest pain to allow for the appearance of the markers in the circulation. For "early markers" such as myoglobin or CK-MB isoforms, this is typically 1 to 4 hours, whereas markers such as CK-MB or the troponins require up to 6 to 8 hours before they can be detected.[15]

Proper application of diagnostic modalities to patients with chest discomfort requires an understanding of pretest likelihoods and how they are applied. Bayes' theorem is a mathematical rule that relates the probability of disease in a patient before the test (pretest probability) to the probability of disease after the test is completed (post-test probability). To understand the application of Bayes' theorem to patient care, a few examples will suffice. If an individual is highly unlikely to have a disease (ie, a healthy young man who has blood drawn for an insurance physical), and then a test is performed (accidental measurement of a pregnancy test), even if the test is "positive" the individual is still highly unlikely to actually be pregnant. In other words, the result is most likely a false-positive result. The post-test likelihood is low due to the extremely low pretest likelihood. Conversely, if an individual presents to the ED with heavy substernal chest pain of 12 hours' duration, has multiple risk factors for coronary artery disease, and manifests new ST-segment elevation on the ECG, a "normal measurement" of a marker of myocardial injury will not exclude the possibility of an AMI. The odds still favor such a diagnosis due to the high pretest likelihood that the individual is having an AMI. A proper understanding of Bayes' theorem can facilitate resolution of these uncommon situations. More importantly, Bayes' theorem allows us to predict that the diagnostic use of a test will be greatest when applied to individuals who have an intermediate likelihood of disease. In these individuals, the results of the test will reveal the greatest difference between pretest and post-test likelihood ratios. Although Bayes' theorem is a valuable tool for the evaluation of diagnostic results in populations, it must be appreciated that individual patients often have important elements that cannot be calculated with robust certainty.

A protocol should be in place that delineates the assessment and management of patients who present with chest discomfort or anginal equivalent.[16] Protocol-driven chest pain centers have been found to be cost efficient and to shorten the time to diagnosis while minimizing unnecessary admissions and inappropriate discharges.[17,18] A 12-lead ECG should be done within 5 to 10 minutes of the patient's arrival to an ED.[3] Continuous cardiac monitoring must be implemented to observe for potentially life-threatening arrhythmias. Serial ECGs improve the sensitivity for detection of ischemia. The measurement of continuous ST-segment monitoring has also been found to significantly improve sensitivity, especially for early risk-factor stratification in patients with nondiagnostic ECGs.[19,20] In the setting of ischemia, the ECG may show ST elevation greater than 1 mm above baseline, T-wave inversion, tall peaked T waves, or it may be completely normal. In patients with AMI, the ECG will often show ST elevation greater than 1 mm in the affected leads. The ECG may also show hyperacute T waves and ST-segment elevation in two or more contiguous affected leads, development of Q waves that are greater than 25% of the size of the QRS and 0.04 seconds wide, deep ST or T-wave abnormalities, or nondiagnostic changes.[21] Although often thought to represent irreversible necrosis, Q waves may develop within 2 hours of thrombus formation.

Many individuals with ACS, however, will have nondiagnostic ECGs because of both the limitations of the ECGs and the labile nature of the underlying pathologic condition. Patients with conduction system disease, left ventricular hypertrophy, drug therapy such as digoxin, or patients with posterior ischemia/infarction are even more likely to have nondiagnostic ECGs.

Accordingly, measurements of markers for myocardial necrosis are commonly used. Numerous biochemical markers have been evaluated for the detection of myocardial ischemia or myocardial cellular necrosis. Many are described in greater detail in this monograph. The ideal marker does not exist, but desirable characteristics include high protein concentrations within the myocardium, low or absent concentration in noncardiac tissue, persistence in circulation, rapid release into the circulation after cell injury, and the ability to develop a reliable and rapid analytic system. For optimal specificity, there should not be significant amounts present in other healthy or diseased tissues. Some "cardiac" proteins can be produced in noncardiac tissue during fetal development, allowing for future re-expression if tissue injury (typically of striated skeletal muscles) occurs. For example, the B chain of CK is produced in large amounts in healthy fetal skeletal muscle cells. Levels of CK-MB are therefore quite high in skeletal muscle tissue in the fetus. The production of the B chain is then downregulated during development, with a rapid fall in the production and presence of CK-MB in healthy skeletal muscle. However, re-expression of the B chain occurs in adult skeletal muscle cells after injury, resulting in a fall in the specificity for the detection of cardiac injury, ie, injury to skeletal muscle that contains CK-MB will cause a "false-positive" rise in the level of circulating CK-MB.

Certain biological and cellular aspects of markers affect their application to diagnostic performance. Smaller molecules tend to be released more rapidly, are detected more quickly, and persist in circulation for shorter periods of time. Cytoplasmic markers (CK, CK-MB, myoglobin, and fatty acid binding protein) are released more rapidly than are structural proteins (such as myosin heavy chains). Structural proteins are released more slowly, but for longer periods of time. Some markers of myocardial injury are found in more than one intracellular pool. For example, small amounts (3% to 7%) of troponin proteins are believed to be present in the cytoplasm/cytoplasmic matrix, whereas the rest is complexed within the contractile apparatus. Thus, the small amount present in the cytoplasm is released rapidly while continuing liberation of the remaining troponin protein is slowly released from the contractile apparatus over a period of days.

The half-life of a particular protein may be quite different from its diagnostic window. For example, the circulating half-life of troponin in the healthy state appears to be quite short (approximately 30 minutes), yet the diagnostic window for troponin I is usually 5 to 7 days, whereas that of troponin T is up to 10 days. This disparity appears to be due to the continued release of troponins from the contractile apparatus. The actual duration in any one patient also will be determined, at least in part, by the size of the infarction. Smaller degrees of myocardial injury will result in correspondingly shorter durations of elevation. The total amount of a marker protein released into circulation as well as its rate of clearance may vary depending on the degree of perfusion. This will complicate assessments of infarct size but may allow for a determination of successful reperfusion after thrombolytic therapy or earlier detection of cardiac injury. Finally, there can be additional modification or degradation of the protein that can occur both within the cell and after it is released into the circulation.[22] This post-translational modification or postrelease degradation can at times be used for diagnostic purposes. However, assays that are highly specific for the native protein may lose sensitivity for detection of the protein once it begins to undergo degradation.

Until recently, measurement of total CK, CK-MB, and lactate dehydrogenase were the standard diagnostic tests used.[15] The limitations of these assays led to the acceptance of improved analytes. Currently, when patients present to the emergency room with chest pain, they will most likely have blood drawn to measure the level of cardiac troponins (either I or T). Samples should be obtained in a serial fashion, with up to three or more samples obtained over a 6- to 12-hour period. Many laboratories still offer measurement of CK-MB in addition to troponin, especially on the first sample. There is a suggestion that levels of CK-MB may rise sooner than levels of troponin rise, at least in some patients. Troponins have been widely accepted in a short period of time, due to both their superior cardiac specificity and their improved sensitivity. Indeed, measurement of cardiac troponins has allowed us to detect smaller degrees of cardiac injury than previously possible.[23,24] Thus, patients who in the past would have been labeled as having unstable angina without injury are now found to have small degrees of cardiac injury. Such patients are frequently said to have "minor myocardial necrosis," in part to designate the minimal degree of injury, as well as to separate this group from those who have classically been said to have "myocardial infarction." However, it should be understood that such patients are still at significantly increased risk when compared with patients in whom no elevation of troponin is present.[4,23] This risk appears to correlate closely with the peak level of troponin as well as with the absence or presence of detectable troponin in the admission sample.

The above discussion applies to currently available biomarkers such as CK-MB or troponin that are felt to be predominantly or exclusively released only after myocardial cell necrosis (rather than after periods of ischemia). Some studies suggest that very low levels of these proteins (especially troponins) may be detectable in the circulation after cellular ischemia occurs, but this issue is not yet resolved. Also, current assay platforms are not designed for the requisite analytic sensitivity for the detection of ischemia to be a practical consideration. Issues noted earlier, such as the desirability of high concentration of biomarkers in myocardium, may not apply to settings of ischemia since many potential analytes are extracardiac in origin. The term ischemia can be applied to different pathophysiological situations that will be very important in the application of blood-based diagnostic markers. Thus, a marker developed for the detection of ischemia may not accurately detect ischemia in all clinical situations, depending on the active pathologic process. Having a marker or markers that could detect ischemia would have a profound clinical impact since in most patients who present with chest pain, detection of ischemia is the pertinent clinical question.

Finally, if the patient has been "ruled out" for myocardial injury, then an evaluation for the presence of inducible cardiac ischemia is indicated. Exercise stress tests form the cornerstone for the diagnosis of inducible ischemia. Exercise stress tests help not only to screen for latent coronary artery disease but also to determine the degree and extent of pathology, to define functional capacity to facilitate the formulation of an exercise prescription, to determine prognosis, and to help assess the effectiveness of medical or surgical therapy. Numerous stress test protocols have been successfully applied to patients. In addition to treadmill testing, pharmacologic stress tests via dobutamine, adenosine, or dipyridamole are all used successfully. The sensitivity and specificity for the detection of coronary artery disease can be improved by the addition of ancillary imaging modalities such as nuclear perfusion imaging (typically thallium or sestamibi) or serial echocardiographic testing. Although used in outpatient settings for decades, exercise stress testing is now increasingly used for the early detection of myocardial ischemia in the ED setting. Patients who have no evidence of ischemia can be discharged home, whereas those with positive stress evaluations should be admitted for further evaluation. Evaluation of coronary artery anatomy with coronary angiography will be indicated in many such patients. Performing the stress test in the ED confers significant cost advantages compared with routine hospital admission, but requires the ability to perform the test in a timely fashion during weekends and at night.

Although stress evaluation for the detection of inducible ischemia is reasonably accurate, there are several limitations to its routine application in all patients. First, some individuals have false-positive studies. False positives can be more prevalent in certain patient subgroups. Second, reliance on stress testing for the routine detection of inducible ischemia can be a labor-intensive endeavor, especially if stress testing is provided at night and on weekends. If testing is only available during weekdays, then patients who present with chest pain must either be held until a stress evaluation is performed, thereby increasing cost, or be discharged to have their stress evaluation as an outpatient, which increases risk. Finally, stress evaluations do not detect vulnerability to plaque rupture, which is a key determination of risk. Indeed, there is substantial evidence that many factors independent of the degree of epicardial artery stenosis are important in future cardiac risk. A patient with a vulnerable atherosclerotic plaque that is not flow limiting may be at substantial short-term risk and yet have no abnormality detected on any stress test. Although the use of stress evaluations for the detection of inducible ischemia and risk assessment is the norm, there is room for improvement and an opportunity for diagnostic strategies that can be applied to the detection of ischemia and risk in individuals with ACS.

The evaluation of patients who present with chest pain is currently a challenging problem. Conceptually, we would want to rapidly identify those patients who have an acute coronary artery syndrome and be able to offer the appropriate therapy to prevent further damage while avoiding unnecessary diagnostic testing in those patients who do not have a cardiac cause of their chest pain. The currently available diagnostic tools cannot readily provide this information; the area of greatest deficit is in the ability to detect myocardial ischemia. Diagnostic modalities that could allow for the prompt detection of plaque rupture, downstream cellular ischemia, and ongoing myocardial cellular necrosis, and that could define the pathophysiological state of the coronary artery would significantly improve and augment our ever-increasing therapeutic armamentarium.

References

1. American Heart Association. *Advanced Cardiac Life Support.* Dallas, TX: American Heart Association; 1997.
2. Pope JH, Aufenheide TP, Ruthazer R, et al. Missed diagnosis of acute cardiac ischemia in the emergency department. *N Engl J Med* 2000;342: 1163–1170.
3. Ryan TJ, Anderson JL, Antman EM, et al. ACC/AHA guidelines for the management of patients with acute myocardial infarction: A report of the American College of Cardiology/American Heart Association Task Force on Practice Guidelines (Committee on Management of Acute Myocardial Infarction). *J Am Coll Cardiol* 1999;28:1328–1428.
4. Zaacks SM, Liebson PR, Calvin JE, et al. Unstable angina and non-Q wave myocardial infarction: Does the clinical diagnosis have therapeutic implications? *J Am Coll Cardiol* 1999;33:107–118.
5. Fuster V. *The Vulnerable Atherosclerotic Plaque: Understanding, Identification, and Modification.* Armonk, NY: Futura Publishing Company, Inc.; 1999.
6. Braunwald E. Unstable angina: An etiologic approach to management. *Circulation* 1998;98:2219–2222.
7. Braunwald E. Recognition and management of patients with acute myocardial infraction. In Goldman L, Braunwald E (eds): *Primary Cardiology.* Philadelphia: W.B. Saunders; 1998:257–283.
8. Dracup KA, Cannon CP. Combination treatment strategies for management of acute myocardial infarction. *Crit Care Nurse Suppl* 1999:3–15.
9. Liuzzo G, Biasucci LM, Gallimore JR, et al. Enhanced inflammatory response in patients with preinfarction unstable angina. *J Am Coll Cardiol* 1999;34:1696–1703.
10. Terres W, Kummel P, Sudrow A, et al. Enhanced coagulation activation in troponin T-positive unstable angina patients. *Am Heart J* 1998;135:281–286.
11. Theroux P, Kuster V. Acute coronary syndromes: Unstable angina and non Q wave MI. *Circulation* 1998;97(1):1195–1206.
12. Casey K, Bedker DL, Roussel-McElmeel P. Myocardial infarction: Review of clinical trials and treatment strategies. *Crit Care Nurse* 1998;18(2):39–52.
13. Hamm CW, Goldman BU, Heeschen C, et al. Emergency room triage of patients with acute chest pain by means of rapid testing for cardiac troponin T or troponin I. *N Engl J Med* 1997;337:1648–1653.
14. Hillis GS, Zhao N, Taggert P, et al. Utility of cardiac troponin I, creatine kinase-MB, myosin light chain 1, and myoglobin in the early in-hospital triage of "high risk" patients with chest pain. *Heart* 1999;82:614–620.

15. Dati F, Panteghini M, Apple FS, et al. Proposals from the IFCC Committee on Standardization of Markers of Cardiac Damage (C-SMCD): Strategies and concepts on standardization of cardiac marker assays. *Scand J Clin Invest* 1999;59(suppl 230):113–123.
16. Farkouh ME, Smars PA, Reeder GS, et al. A clinical trial of a chest-pain observation unit for patients with unstable angina. *N Engl J Med* 1998;339: 1882–1888.
17. Lange RA, Cigarroa JE, Hillis LD. Thrombolysis or primary PTCA for acute myocardial infarction. *ACC Educational Highlights* 1997;12(3):104.
18. Ornato JP. Chest pain emergency centers: Improving acute myocardial infarction care. *Clin Cardiol* 1999;22(suppl IV):IV3-IV9.
19. Jernberg T, Lindahl B, Wallentin L. ST-segment monitoring with continuous 12-lead ECG improves early risk factor stratification in patients with chest pain and ECG nondiagnostic of acute myocardial infarction. *J Am Coll Cardiol* 1999;34:1413–1419.
20. Holmvang L, Luscher MS, Clemmensen P, et al. Very early risk stratification using combined ECG and biochemical assessment in patients with unstable coronary artery disease (A Thrombin Inhibition in Myocardial Ischemia [TRIM] Substudy). *Circulation* 1998;98:2004–2009.
21. Miracle VA, Sims JM. Making sense of the 12 lead ECG. *Nursing* 1999;29(7): 34–39.
22. McDonough JL, Arrell DK, Van Eyck JE. Troponin I degradation and covalent complex formation accompanies myocardial ischemia/reperfusion injury. *Circ Res* 1999;84:9–20.
23. Antman EM, Tanasijevic MJ, Thompson B, et al. Cardiac-specific troponin I levels to predict the risk of mortality in patients with acute coronary syndromes. *N Engl J Med* 1996;335:1342–1349.
24. Ohman EM, Armstrong PW, White HD, et al. Risk stratification with point-of-care cardiac troponin T test in acute myocardial infarction. *Am J Cardiol* 1999;84:1281–1286.

Chapter 7

WHO Criteria: *Where Do We Go from Here?*

Allan S. Jaffe, MD

Introduction

The initial World Health Organization (WHO) criteria did not include biomarkers of cardiac injury. The initial technical report in 1959 emphasized the electrocardiogram and failed to mention blood tests of any kind.[1] This is not unexpected since it was not until 1954 that the first reports of the use of serum transaminases to facilitate the diagnosis of acute myocardial infarction (AMI) were published by Karmen et al.[2] This initial observation of increases in serum glutamic-oxaloacetic transaminase proximate to the time of the acute event began the field of biomarkers for the diagnosis of MI. The 1970 WHO criteria took notice of this advance and indicated that MI could be diagnosed predicated on increases in marker proteins (they used the term enzymes) that were measured locally.[3] Subsequently, in 1979 when plans were being made for the MONICA project, the criteria were changed again, and definite MI was defined at that time as elevations in enzymes that were two times greater than the upper limit of the reference range.[4] It was suggested that lower elevations would be considered indefinite or indeterminate.

These criteria were promulgated predominately for epidemiological reasons. The reports from the project make it clear that the WHO was interested in making sure that they could characterize definite events and that they therefore chose to emphasize the specificity of diagnosis to a greater extent than they did the sensitivity.[5,6] The writing of Dr. Tunstall-Pedoe emphasized his concern that as we added criteria

Dr. Jaffe is a consultant to Dade Behring, which has also supported research in the author's laboratory.

From: Adams JE III, Apple FS, Jaffe AS, Wu AHB (eds). *Markers in Cardiology: Current and Future Clinical Applications.* Armonk, NY: Futura Publishing Company, Inc.; © 2001.

based on marker proteins, a more sensitive set of criteria were being promulgated and that the incidence of acute infarction might be altered. He agreed that it was important to have more definitive information about the group that was considered "indeterminate" or "indefinite." These problems persist today and are now amplified by the development of new markers such as cardiac troponins.

Troponin Markers

The purpose of this chapter is not to review the comprehensive data available concerning the cardiac troponins T and I (cTnT and cTnI). However, it should be clear that both cTnI and cTnT manifest increased, if not unique, specificity for the heart. In initial basic science studies, cTnI has not been found to be expressed anywhere except in cardiac tissue, during either adult life or neonatal development.[7] Thus, diseased skeletal muscle, a site where the muscle proteins that are expressed during neonatal development are often re-expressed in response to injury, does not contain cTnI.[8] This observation made by Bodor and colleagues[8] in patients with polydermatomyositis and Duchenne's muscular dystrophy strongly supports the notion that not even small amounts of cTnI are expressed in skeletal muscle during neonatal development. Similar observations have not been made for every tissue, but thus far no definite exceptions have been reported.

The issue of specificity of cTnT, however, has been more complex. Early on, the issue was confounded by the fact that there was a nonspecific tag antibody in the initial assays that caused some cross-reactivity with skeletal muscle troponin T.[9] Once this cross-reactivity was corrected, the number of elevations of troponin T went down substantially in patients with skeletal muscle injury.[10] If one uses polyclonal serum to the cTnT in an attempt to probe damaged skeletal muscle, cardiac isoforms are observed.[11] However, the specific two antibodies now used in the present cTnT assay do not in combination capture and label the cardiac isoforms of cTnT found in skeletal muscle. Thus, elevations of cTnT detected by the present iteration of the assay should have unique specificity for the heart, equivalent to that of cTnI.[12,13] Only one case has been reported that suggests ectopic production of cTnT, and that report is from a patient who received cardiotoxic therapy that, rather than ectopic production, may well have been the cause of the elevation.[14]

These observations in isolated skeletal muscles have been confirmed in a multitude of clinical studies.[10,15–18] For example, in our own experience, increases in cTnI were seen only in patients who had perioperative evidence of myocardial injury indicated by echocardiography, and increases were not observed in the absence of such changes.[15] This was in sharp contrast to increases in the MB isoenzyme of creatine kinase (CK-MB), which were common, most likely due to release from skeletal muscle. The use of percentage criteria in this circumstance, as in others where there is conjoint skeletal muscle and cardiac injury, resulted in improved specificity but at the cost of sensitivity.[15,16,19]

The troponin markers are not only more specific but also more sensitive[20–22] for cardiac damage. Elevations persist in plasma for a longer period of time than for CK-MB or lactate dehydrogenase.[22–24] Thus, the greater incidence of increases in cardiac troponins in the absence of increases in CK-MB in patients with unstable angina is well reported and is in the range of 33%.[25] If one evaluates patients with ischemic chest pain who have minimal elevations of total CK, a large percentage of these patients have elevations in troponin without substantial elevation in CK-MB. Furthermore, percentage increments in CK-MB are substantially less than the percentage increments in troponin.[20]

This sensitivity combined with the fact that troponin elevations persist in plasma for a prolonged period time because the pool that is bound to the contractile apparatus is degraded and released slowly[24] leads to a substantially greater number of increases in troponin than in CK-MB.

This heightened sensitivity has led to questions about whether troponins may "leak out of cells" in the absence of irreversible injury. No data exist to answer this question thus far. However, unpublished information suggests that coronary sinus samples do not increase in response to the induction of ischemia (H. Katus, unpublished observations, University of Lübeck, 1994 to 1995), and that in models of vital exhaustion where increases in troponins are observed, elevations of troponins are invariably associated with pathologic changes of necrosis.[26] One pathologic study documenting necrosis, when increases in cTnI were observed clinically in the absence of other findings, suggested that elevations are associated with the evidence of necrosis.[27] Experimental data in this area are available only for total CK and emphasize the difficulty of determining whether troponins can leak out of cells. In a model of local ischemia induced by balloon occlusion, animals invariably had histologic evidence of necrosis if occlusion occurred for more than 20 minutes, and total CK was increased in the circulation.[28] In two animals with occlusions for only 15 minutes, increases in CK and plasma were observed. Comprehensive analysis using electron microscopy found necrosis in the ischemic beds of both of these animals. This finding suggests that as far as can be ascertained, proteins such as CK, which are found in the cytosol, are released in response to irreversible injury only. The troponins, though substantially smaller in molecular weight than CK, appear to be released as complexes that are closer in weight to CK.[29] Thus, by analogy, it is likely, although not proven, that increases in the troponins are indicative of irreversible cardiac injury. The nature of what is released, however, may vary depending on the degree of ischemia, as recently reported by McDonough and colleagues.[30]

Recommendations of Others

Because of the increased sensitivity and specificity of the troponins, several groups have recommended that the troponins replace CK-MB as the marker of choice for the diagnosis of AMI.[31,32] I agree. There are, however, two groups of patients who present diagnostic difficulties when one makes this change. The first group consists of patients who have elevated troponin values but normal CK-MB values and have a clinical presentation suggestive of acute coronary syndromes. The second group consists of patients who have elevated troponin values and normal CK-MB values but have a clinical presentation that is not suggestive of acute coronary syndromes.

Patients with Probable Acute Coronary Syndromes

The predictive accuracy of the value of troponins in patients with acute coronary syndromes is well reported.[25] Prognosis is adverse when evidence of cardiac injury determined by troponin is present even in the absence of increases in other marker proteins. It is important to note, however, that if one changes the cutpoints for other markers and begins to look at subreference limit changes in all conventional markers such as isoforms, myoglobin, as well as the troponins, minor increases are observed that have prognostic value.[33] Thus, increases are not a function of specific markers but relate to the more sensitive detection of cardiac injury.

The National Academy of Clinical Biochemistry has suggested that two cutpoints for the troponins be used, a value indicative of AMI and a second value indicative of cardiac injury but in which AMI is not diagnosed.[32] This is a controversial recommendation and has both pros and cons (Table 1).

In favor of this recommendation is the concept that one can continue to use the existing paradigms for the clinical evaluation of patients. Clinicians have been happy over the years with being able to distinguish patients with elevations of CK-MB as those with acute in-

Table 1

Pros and Cons for Two Cutpoints for Troponin Values in Patients with ACS

PROS
1. Allows continuation of a well developed and effective clinical paradigm.
2. Preserves ability to tract epidemiology based on consistent criteria.
3. Avoids insurance problems for some patients.
4. Allows more time to develop better and more data upon which to ensure the specificity of the troponins.
5. Allows more time to develop better data upon which to base risk stratification.
6. Allows time for assay problems to be solved.

CONS
1. Science suggests that increased sensitivity is involved, not differences in mechanism.
2. Prognosis appears to be similar in many patients with non-Q-wave infarction.
3. Delay in implementing new guidelines will make it harder eventually to solve the problem.
4. Sufficient data concerning specificity and risk statification exist.
5. Many of the assay problems are resolved or should be soon.

ACS = acute coronary syndrome.

farction, and those without elevations of CK-MB as having unstable angina. Clinicians have developed clinical paradigms for treatment, many of which they would like to continue using. In addition, such paradigms allow one to preserve the ability to track the incidence of MI epidemiologically. Furthermore, by not changing criteria, one avoids societal problems related to insurance as well as to employment for many patients. One could argue that a delay in making changes would also allow one to ensure better data and better risk stratification and permit assay problems to be solved before changing the diagnosis of acute infarction.

Against this recommendation, the reality appears to be that troponin has increased sensitivity of detection of cardiac injury, but not of the mechanism of injury. Prognosis appears similar in patients with elevated troponins and "clinical unstable angina" and those with non-Q-wave infarction.[34] Thus, sufficient data exist today to allow implementation of these new guidelines. Given there are increasing data that risk stratification can be done sensibly using conjoint measures of troponin elevations and the electrocardiogram,[25] the time is probably right to begin to make the change in paradigm. Indeed, many of the assay problems are being addressed. These assay problems already have been solved for the cTnT assay. Thus, the need for two cutpoints in patients with acute coronary syndromes is related more to the comfort of physicians rather than to the science.

In regard to patients who do not have acute coronary syndromes, these elevations of troponin often are problematic (Table 2). It is difficult to know how to care for these patients aside from labeling them and doing unnecessary work-ups that may or may not be cost effective but may be deemed essential. It has long been known, however, that elevations of markers of cardiac injury do not indicate the mechanism of damage, and it is clear that the mechanism of damage may or may not be prognostically important. For example, there may be reasons for transient elevations, such as viral infections. Thus, finding the appropriate cause may be essential in individual patients. It is now clear that in many diseases that are likely associated with cardiac damage such as congestive heart failure,[35,36] hypertension and toxic insults with chemotherapeutic agents[37,38] may well cause cardiac injury and be confused with AMI. It is not logical to deal with these issues when the troponin levels are high but ignore them when the troponin levels are marginal. One example of progress in this area was presented a few years ago by Lauer and colleagues.[39] Patients who presented with chest

Table 2

Pros and Cons for Two Cutpoints for Troponin Values in Patients without Apparent ACS

PROS

1. The cause of elevations is often hard to determine, and thus appropriate acute and long-term management is unknown.
2. Patients will be labeled, and that will have effects both medically and perhaps in regard to insurance and employment.
3. Unnecessary work-ups will be costly but essential.

CONS

1. Elevated markers say nothing about the mechanism of damage, but only that damage is present.
2. The presence of cardiac damage may or may not be prognostically important, depending on the mechanism of the damage.
3. There are many other disease mechanisms that could lead to cardiac damage, eg, CHF and hypertension.
4. We may be underdiagnosing certain disease entities, for example, toxic insults and myocarditis.
5. Time waits for nobody

ACS = acute coronary syndrome; CHF = congestive heart failure.

Table 3

A Modest Proposal

- Troponin markers are the tests of choice for the detection or exclusion of cardiac injury.
- Detectable elevations are abnormal.
- Elevations of troponin define the presence of cardiac injury, not the mechanism.
- Damage due to "acute ischemic heart disease" should be diagnosed when there are elevations of troponin in a clinical situation that suggests that the mechanism of injury is ischemia.
- Treatment and prognosis are predicated on the nature of the presentation and the extent of elevations.
- Elevations of troponin not attributable to "acute ischemic heart disease" will require alternative diagnoses and management predicated on the mechanism of injury. At times, follow-up may be all that is necessary.

pain, often with electrocardiographic changes and elevated cTnT, underwent biopsy to look for myocarditis. Very few patients met criteria for myocarditis via the Dallas criteria that required increased lymphocytic infiltration and evidence of necrosis. However, when immunohistochemistry was done, 28 of the 29 patients with increases in cTnT had abnormal accumulations of lymphocytes, suggesting that it was the sampling bias that missed the evidence of necrosis and therefore confused the diagnosis. Troponin elevations were more sensitive and therefore facilitated the evaluation of myocarditis. Only with the ability to make more precise diagnoses will the field advance. This is but one example of what may be a large number of additional disease entities such as microvascular disease and problems associated with renal failure that we will misdiagnose unless we embrace a new standard for the definition of cardiac injury. Thus, despite the discomfort, this author would suggest a modest proposal (Table 3).

Recommendations

Troponin markers should be the test of choice for the detection of possible cardiac injury. Detectable elevations should be considered abnormal. To make the use of troponin markers the gold standard, laboratorians and manufacturers must deal with the problems of variability, of assays that have not always been as precise as they should be, and of increases related to analytic interferences.[31,32] Once elevations have been detected, they define the presence of cardiac injury but not the mechanism. Those due to acute ischemic heart disease should be diagnosed when the elevations of troponin occur in a situation that the clinician defines as due to ischemia. Treatment and prognosis at that point is predicated on the nature of the clinical presentation and the extent of the elevations. However, elevations that occur in the absence of a clinically defined acute ischemic insult should not be labeled AMI. Thus, an elevation does not mandate the diagnosis of MI; it simply mandates the fact that cardiac injury has occurred. Other work-up may be needed to define the cause of the elevation, its prognostic significance, and the appropriate therapy.

Other Issues Related to AMI Diagnosis

Some groups have suggested that a second rapidly rising marker be used in patients with possible acute ischemic heart disease. This is a reasonable approach if the use of such values will facilitate management. It may be that in some patients the use of a second rapidly rising marker will allow for earlier initiation of aggressive therapy or earlier treadmill testing to allow for earlier discharge. However, unless these values result in changes in patient management, the use of rapid acting markers is not advocated.[40] In terms of which marker might be best, recent control information finds a slightly greater percentage of patients diagnosed early with isoforms than with myoglobin, although the differences are not statistically significant.[33] Fatty acid binding protein could also enter this arena and compete successfully for use in this circumstance.[41] Finally, recent data suggest that if one uses the low cutoff for the troponins, troponin markers can be definitive as early as 6 hours, and this approach may well, in the long run, supplant the others.[42]

Two other issues should be addressed. One relates to the use of point of-care-testing (POCT). I agree with the recommendations of others[31,32]: POCT should be implemented whenever a laboratory is unable to provide sufficiently rapid turnaround time. Turnaround times for most laboratories should be within 1 hour at the worst, and in the absence of the ability to provide such service, it is justified to use POCT apparatuses in my view. In the long run, improvements in POCT technology will facilitate its use, and bedside tests will be so facile and so accurate that laboratory measurements will not be needed. Nonetheless, at present, POCT has a place in those settings in which laboratory values are not readily available.

Finally, if one defines increases in marker proteins as indicative of cardiac injury, one must address the issue of the interventional cardiologist. Increases of troponin in the interventional setting are extremely common and often lead to confusion about both the etiologic mechanism and the prognosis.[43,44] My suggestion would be that because such increases in troponin occur in an ischemic milieu, they should be considered indicative of infarction regardless of their magnitude. However, occlusion and reperfusion substantially increase the release ratio of CK[45] and likely all markers, and by doing so increase the sensitivity of detection. Thus, terms such as minimal or minor should be used to describe such insults. Finally, it is unfair to use these endpoints in the same way that spontaneous infarctions are used for clinical trials or for epidemiological or reimbursement purposes. The management of such individuals depends on the adequacy of the procedure as well as the magnitude of the elevation. Thus, separating out elevations of troponins in the interventional setting and making it clear that the use of measurement of troponins represents the most sensitive levels of detection seems the most reasonable approach.

Conclusion

This is an exciting time, when the science must dictate our new paradigm. The adoption of a new paradigm is uncomfortable for clinicians who in the past could ignore CK-MB elevations because there was adequate justification and knowledge of false-positive increases. The troponins remove such flexibility and therefore test us all clinically. We are up to the challenge.

References

1. WHO Expert Committee on Cardiovascular Disease and Hypertension. Hypertension and coronary heart disease: Classification and criteria for epidemiological studies. World Health Organization Technical Support Series No 168:3–28, 1959.
2. Karmen A, Wroblewski F, Ladue J. Transaminase activity in human blood. *J Clin Invest* 1955;34:126–133.
3. Anonymous. The pathological diagnosis of acute myocardial infarction: Preliminary results of a WHO cooperative study. Bulletin of the World Health Organization. 1973;48:23–25.
4. Tunstall-Pedoe H, Kuulasmaa K, Amouyel P, et al. Myocardial infarction and coronary deaths in the World Health Organization MONICA Project. Registration procedures, event rates, and case-fatality rates in 38 populations from 21 countries in four continents. *Circulation* 1994;90:583–612.
5. Tunstall-Pedoe H. Perspectives on trends in mortality and case fatality from coronary heart attacks: The need for a better definition of acute myocardial infarction. *Heart* 1998;80:121–126.
6. Martin CA, Hobbs MS, Armstrong BK. Measuring the incidence of acute

myocardial infarction: The problem of possible acute myocardial infarction. *Acta Medica Scand Suppl* 1988;728:40–47.
7. Toyota N, Shimada Y. Differentiation of troponin in cardiac and skeletal muscles in chicken embryos as studied by immunofluorescence microscopy. *J Cell Biol* 1981;91:497–504.
8. Bodor GS, Survant L, Voss EM, et al. Cardiac troponin T composition in normal and regenerating human skeletal muscle. *Clin Chem* 1997;43: 421–423.
9. Katus HA, Looser S, Hallermayer K, et al. Development and in vitro characterization of a new immunoassay of cardiac troponin T. *Clin Chem* 1992; 38:386–393.
10. Muller-Bardorff M, Hallermayer K, Schroder A, et al. Improved troponin T ELISA specific for cardiac troponin T isoform: Assay development and analytical and clinical validation. *Clin Chem* 1997;43:458–466.
11. Bodor GS, Survant L, Voss EM, et al. Cardiac troponin T composition in normal and regenerating human skeletal muscle. *Clin Chem* 1997;43: 476–484.
12. Ricchiuti V, Voss EM, Ney A, et al. Cardiac troponin T isoforms expressed in renal diseased skeletal muscle will not cause false-positive results by the second generation cardiac troponin T assay by Boehringer Mannheim. *Clin Chem* 1998;44:1919–1924.
13. Haller C, Zehelein J, Remppis A, et al. Cardiac troponin T in patients with end-stage renal disease: Absence of expression in truncal skeletal muscle. *Clin Chem* 1998;44:930–938.
14. Isotalo PA, Greenway DC, Donnelly JG. Metastatic alveolar rhabdomyosarcoma with increased serum creatine kinase MB and cardiac troponin T and normal cardiac troponin I. *Clin Chem* 1999;45:1576–1578.
15. Adams JE, Sicard G, Allan BT, et al. More accurate diagnosis of perioperative myocardial infarction with measurement of cTnI. *N Engl J Med* 1994; 330:670–674.
16. Adams JE III, Davila-Roman VG, Bessey PQ, et al. Improved detection of cardiac contusion with cardiac troponin I. *Am Heart J* 1996;131:308–312.
17. Baum H, Braun S, Gerhardt W, et al. Multicenter evaluation of a second-generation assay for cardiac troponin T. *Clin Chem* 1997;43:1877–1884.
18. Mair P, Mair J, Koller J, et al. Cardiac troponin T release in multiply injured patients. *Injury* 1995;26:439–443.
19. Adams JE, Bodor GS, Davila Roman BG, et al. Cardiac troponin I: A marker with high specificity for cardiac injury. *Circulation* 1993;88:101–106.
20. Apple FS, Falahati A, Paulsen PR, et al. Improved detection of minor ischemic myocardial injury with measurement of serum cardiac troponin I. *Clin Chem* 1997;43:2047–2051.
21. Adams J, Scheckman KB, Landt J, et al. Comparable detection of acute myocardial infarction by creatine kinase MB isoenzyme and cardiac troponin I. *Clin Chem* 1994;4017:1291–1295.
22. Katus HA, Remppis A, Neumann FJ, et al. Diagnostic efficiency of troponin T measurements in acute myocardial infarction. *Circulation* 1991;83: 902–912.
23. Katus HA, Remppis A, Scheffold T, et al. Intracellular compartmentation of cardiac troponin T and its release kinetics in patients with reperfused and nonreperfused myocardial infarction. *Am J Cardiol* 1991;67:1360–1367.
24. Jaffe AS, Landt J, Pavvin CA, et al. Comparative sensitivity of cardiac tropo-

nin I and LD isoenzymes for the diagnosis of acute myocardial infarction. *Clin Chem* 1996;42:1770–1776.
25. Klootwijk P, Hamm C. Acute coronary syndromes: Diagnosis. *Lancet* 1999; 353(suppl II):10–15.
26. Chen YJ, Serfass RC, Mackey-Bojack S, et al. Cardiac troponin T alterations in myocardium and serum of rats following stressful, prolonged intense exercise. *J Appl Physiol* 2000;88:1749–1755.
27. Antman EM, Grudzien C, Mitchell RN, Sacks DB. Detection of unsuspected myocardial necrosis by rapid bedside assay for cardiac troponin T. *Am Heart J* 1997;133:596–598.
28. Ishikawa Y, Saffitz JE, Mealman TL, et al. Reversible myocardial ischemic injury is not associated with increased creatine kinase activity in plasma. *Clin Chem* 1997;43:467–475.
29. Wu AH, Feng YJ, Moore R, et al. Characterization of cardiac troponin subunit release into serum after acute myocardial infarction and comparison of assays for troponin T and I. *Clin Chem* 1998;44(6 Pt. 1):1198–1208.
30. McDonough JL, Arrell DK, Van Eyk JE. Troponin I degradation and covalent complex formation accompanies myocardial ischemia/reperfusion injury. *Circ Res* 1999;84:9–20.
31. Panteghini M, Apple FS, Christenson RH, et al. Proposals from IFCC Committee on Standardization of Markers of Cardiac Damage (C-SMCD): Recommendations on use of biochemical markers of cardiac damage in acute coronary syndromes. *Scand J Clin Lab Invest Suppl* 1999;230:103–112.
32. Wu AH, Apple FS, Gibler WB, et al. National Academy of Clinical Biochemistry Standards of Laboratory Practice: Recommendations for the use of cardiac markers in coronary artery diseases. *Clin Chem* 1999;45:1104–1121.
33. Zimmerman J, Fromm R, Meyer D, et al. Diagnostic marker cooperative study for the diagnosis of myocardial infarction. *Circulation* 1999;99: 1671–1677.
34. Ravkilde J, Nissen H, Horder M, Thygesen K. Independent prognostic value of serum creatine kinase isoenzyme MB mass, cardiac troponin T and myosin light chain levels in suspected acute myocardial infarction. *J Am Coll Cardiol* 1995;25:574–581.
35. Missov E, Mair J. A novel biochemical approach to congestive heart failure: Cardiac troponin T. *Am Heart J* 1999;138(1 Pt. 1):95–99.
36. Missov E, Calzolari C, Pau B. Circulating cardiac troponin I in severe congestive heart failure. *Circulation* 1997;96:2953–2958.
37. Missov E, Calzolari C, Davy JM, et al. Cardiac troponin I in patients with hematologic malignancies. *Coron Artery Dis* 1997;8:537–541.
38. Fink FM, Genser N, Fink C, et al. Cardiac troponin T and creatine kinase MB mass concentrations in children receiving anthracycline chemotherapy. *Med Pediatr Oncol* 1995;25:185–189.
39. Lauer B, Niederau C, Kuhl U, et al. Cardiac troponin T in patients with clinically suspected myocarditis. *J Am Coll Cardiol* 1997;30:1354–1359.
40. Jaffe AS. More rapid biochemical diagnosis of myocardial infarction: Necessary? Prudent? *Clin Chem* 1993;39:1567.
41. Key G, Schreiber A, Feldbrugge R, et al. Multicenter evaluation of an amperometric immunosensor for plasma fatty acid-binding protein: An early marker for acute myocardial infarction. *Clin Biochem* 1999;32:229–231.
42. Hamm CW, Goldmann BU, Heeschen C, et al. Emergency room triage of

patients with acute chest pain by means of rapid testing for cardiac troponin T or troponin I. *N Engl J Med* 1997;337:1648–1653.
43. Califf RM, Abdelmeguid AE, Kuntz RE, et al. Myonecrosis after revascularization procedures. *J Am Coll Cardiol* 1998;31:241–251.
44. Johansen O, Brekke M, Stromme JH, et al. Myocardial damage during percutaneous transluminal coronary angioplasty as evidenced by troponin T measurements. *Eur Heart J* 1998;19:112–117.
45. Vatner SF, Baig H, Manders WT, Maroko PR. Effects of coronary artery reperfusion on myocardial infarct size calculated from creatine kinase. *J Clin Invest* 1978;61:1048–1056.

Chapter 8

An Integrated Diagnostic Approach to the Patient with Chest Pain

Robert L. Jesse, MD, PhD and Michael C. Kontos, MD

Introduction

Chest pain is a common complaint among patients presenting to the emergency department (ED), accounting for 5% to 8% of all ED visits.[1] A majority of these patients are admitted due to diagnostic uncertainty, though most are later determined to have nonischemic causes for their symptoms.[2] Even with this relatively low threshold for admission, 30,000 to 40,000 patients per year in the US are released from EDs and are later found to have acute myocardial infarction (AMI). This is associated with significant morbidity and mortality.[3,4] Some of these patients may have ongoing AMI, though most are likely to have unstable coronary syndromes that only later evolve into an AMI. Unfortunately, they are also likely to have more atypical symptoms and nondiagnostic electrocardiograms (ECGs), making it difficult to achieve an acceptable level of sensitivity using the standard ED evaluation tools.

The history, physical examination, and ECG are simply not adequate to exclude myocardial ischemia in the acute setting. Graff et al. analyzed the results from five large chest pain studies[5–9] and demonstrated an inverse correlation between the proportion of patients admitted and the missed MI rate, ie, the fewer patients admitted, the higher the error rate.[10] As seen in Table 1, if the admission rate is 1/3, the missed MI rate is approximately 10%,[5,6] if 1/2 it is 5%,[7,8] and if 2/3 it is 3%.[9] In the Graff study,[10] 67% of the patients were admitted or "ruled out" in a chest pain evaluation unit, and the MI rate among patients

From: Adams JE III, Apple FS, Jaffe AS, Wu AHB (eds). *Markers in Cardiology: Current and Future Clinical Applications.* Armonk, NY: Futura Publishing Company, Inc.; © 2001.

Table 1

Relationship between the Admission Rates for Chest Pain Patients and the Rates of Missed Acute Myocardial Infarction (AMI)

			Relative Proportions	
Reference	Admit Rate	Missed AMI Rate	Admissions	Missed AMIs
Tierney[5]	30%	13%		
Rouan[6]	33%	10%	1/3	10%
Pueleo[8]	53%	4.6%		
Goldman[7]	59%	4.7%	1/2	5%
Graft[10]	67%*	3.0%†		
Selker[9]	72%	2.8%	2/3	3%

*Includes patients admitted to hospital or to observation; †AMI rate among patients in observation unit.

in the chest pain observation unit was 3%. The error rates do not appear to be a function of the physicians' training. Postgraduate experience improves the ability to identify those patients who are having an acute coronary syndrome (ACS): 88% for postgraduate year 1 (PGY1) versus 93% for those beyond PGY3, but it occurs at the expense of a higher admission rate for non-ACS patients, 34% versus 47%, respectively.[11]

The initial decision at the time of presentation is to distinguish low-risk non-ACS patients from the higher risk ACS patients. Based on history, physical examination, and ECG findings, the best sensitivity for this risk stratification appears to be around 93% to 96%.[7,12,13] How then can we improve sensitivity to accurately identify all high-risk patients and at the same time allow for the safe discharge of the low-risk patients? The answer lies in the integration of new technologies into the evaluation process.

Evaluation of Patients Presenting with Chest Pain

The ACS is a complicated spectrum of events that we categorize through a limited number of clinical descriptors including Q-wave or non-Q-wave MI, ST-segment or non-ST-segment elevation AMI, and unstable angina. This pathophysiological process is a continuum, as is the relative risk distribution across the spectrum of patients presenting with chest pain. Effective risk stratification of ACS patients must involve the detection of events across this continuum, including those that precede myocardial necrosis. This is increasingly important as new therapies become available that can reduce morbidity and mortality for ACS patients in these earlier stages.

The ability to risk stratify patients in a systematic fashion is dependent upon a fundamental understanding of the pathophysiology underlying the ACS. The precipitating event appears to be the disruption of a "vulnerable" atherosclerotic plaque that has been weakened by local inflammation. This leads to adhesion and activation of platelets in response to the exposure of subendothelial components; then, with initiation of the plasma coagulation cascade, it can lead to the formation of an intracoronary thrombus. If the burden may be small with little to no effect on blood flow, then the event is asymptomatic. However, when there is significant reduction in blood flow, the oxygen supply to the myocardium is compromised and the tissue becomes ischemic. When the patient is symptomatic, this is clinically defined as unstable angina. If ischemia persists, the myocardium will die; the detection of the products of necrosis clinically defines AMI.[14]

Improving outcomes in ACS requires an accurate clinical risk assessment and a risk-benefit analysis based on prognosis with some given intervention versus prognosis without it; this must include the probability of success as well as the associated risks. As we integrate new technologies into clinical practice, it is important to define their ability to predict both prognosis and the benefit from performing these interventions.

An initial risk stratification of the suspected ACS patient is based on history, physical examination, and ECG. The history is extremely important, but although characteristics of the pain and associated symptoms such as diaphoresis, vomiting, and shortness of breath may be useful for risk stratification, they are not definitive. The most important factors predictive of an acute event are a history of coronary disease, advanced age, and gender. However, traditional "cardiac risk factors" may be misleading because these are more precisely epidemiological predictors for the development of coronary artery disease over time, and not necessarily predictors of acute events. One exception may be diabetes, which has been shown to be predictive in some but not all studies,[15–17] although the incidence of diabetes for patients with AMI is only approximately 30%. Hypertension, smoking, and family history have a similarly low prevalence among these patients, such that reliance on them for risk stratification may cause the risk to be either overestimated or underestimated, and thus they are of limited use in the ED. However, it is important to appreciate that the history can be just as important by suggesting potential noncardiac causes for the presenting symptoms, and thus it should not be ignored.

Like the history, the physical examination is often more helpful for finding alternative explanations for the symptoms than it is for establishing a definitive cardiac cause. However, signs and symptoms suggestive of impaired left ventricular dysfunction carry important prognostic value; a third heart sound, rales, jugular venous distention, tachycardia, and hypotension are all associated with an increased mortality.[18–20]

The ECG is the best initial objective test for the evaluation of patients with possible ACS.[21] It is rapid, widely available, relatively inexpensive, and provides both diagnostic and prognostic value. ST elevation is the only criterion that identifies AMI patients who will benefit from fibrinolytics. ST depression also identifies a high-risk patient group, though the exact treatment is not always clear.[22,23]

Integration of Technology in the Approach to the Patient with Chest Pain

Inadequacies of the history, physical examination, and ECG are apparent from the rates of missed AMI. The inability to identify this small subset of patients, fueled by the probability of litigation when it is missed, has driven liberal admission policies over the past decade. The obvious need has been to improve the diagnostic and prognostic accuracy of the initial ED evaluation. The integration of new technology into the process can reduce the rate of inadvertent discharges, but at what cost? Increased expenditures incurred by the expanded use of technology must be offset; presumably this can occur through a reduction in unnecessary admissions and litigation costs. To accomplish this, systematic protocols must be used; these should be risk-based, time-dependent, and constructed to meet specific outcomes goals.

Each step in the pathophysiology of the ACS provides opportunity to improve the evaluation process. However, because the ACS is composed of a sequence of events, there are temporal and severity issues that must be considered (Fig. 1). For instance, the ECG has historically been the mainstay of chest pain evaluation but, in fact, can only detect the electrical effects of ischemia. This is a fairly late event and is very much dependent upon the severity of the ischemic insult at the exact time that the recording is performed. The release of biochemical markers of necrosis is a terminal event and is also dependent on both the severity and extent of ischemia. An ideal test for the ACS would detect the process early, ie, before necrosis has occurred, and be relatively independent of the extent of the insult. Progress toward this is being made: numerous tests and assays to detect various components of this process are under active investigation. Brief examples follow.

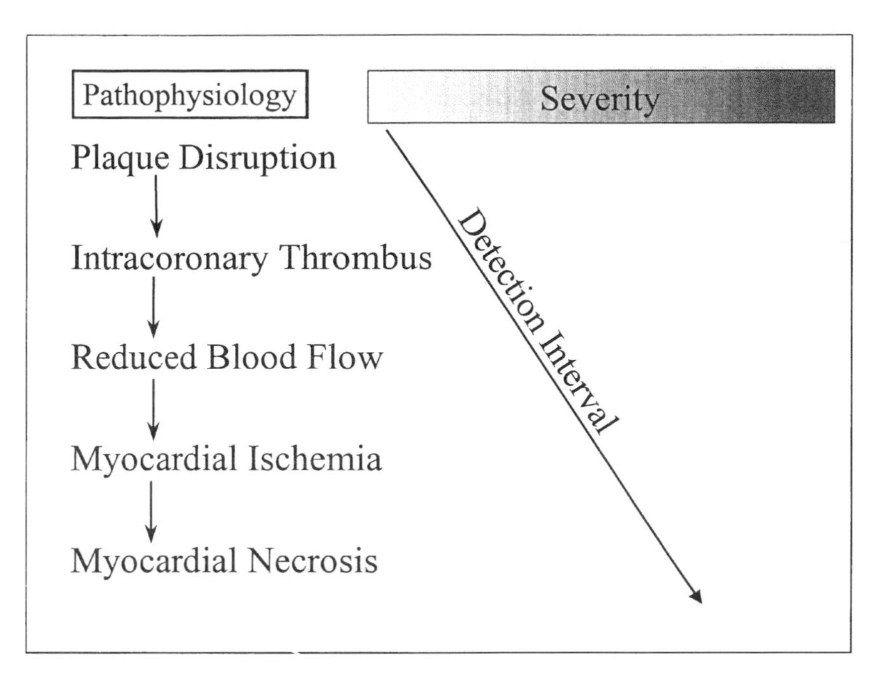

Figure 1. Temporal and severity issues in the pathophysiology of the acute coronary syndrome. The events that comprise the acute coronary syndrome proceed in a stepwise fashion, and are subject to both timing and severity modifications. It is unclear how long a plaque may remain unstable or vulnerable. Although thrombosis may occur rapidly, many patients may experience a stuttering course of symptoms over several days before developing an infarct, suggesting that this is a very dynamic process. It is clear that in the presence of a total coronary occlusion, there is rapid progression to necrosis if there is not rapid restoration of blood flow.

Plaque Disruption

The ruptured plaque is by definition a "vulnerable" plaque, thought to have been weakened due to local inflammation. The cause of this inflammation has not been elucidated, but infection is one possibility that is under scrutiny.[24] The evidence for inflammation provides insight into the underlying pathophysiology and suggests a potential prognostic marker. Liuzzo et al.[25] demonstrated a significant increase in the proportion of patients with elevated C-reactive protein (CRP) among patients having AMI or unstable angina as compared with patients with stable angina. Abdelmouttaleb and colleagues[26] showed a four- to fivefold increase in CRP for patients with recent MI (8.7 mg/L) or unstable angina (11.6 mg/L) relative to controls (2.3 mg/L). Ridker et al.[27] demonstrated that CRP levels at the time of enrollment for volunteers in the Physician's Health Study correlated with the relative risk of subsequent AMI. Patients in the highest quartile had a fourfold risk for AMI over those in the lowest quartile. Further, when randomized to receive aspirin, there was a significant reduction in risk relative to those randomized to receive placebo: the greatest reduction was 56% for those in the highest CRP quartile, but only 14% in the lowest. Studies such as these have raised an interest in using CRP for the initial risk assessment of chest pain patients. An obvious disadvantage is that it is a relatively nonspecific acute phase reactant. However, it does seem to identify patients at high risk as well as those patients who may reduce that risk in response to therapy. If the sensitivity of elevated CRP for ACS patients proves high, normal levels could potentially play an important role in identifying chest pain patients at lower risk.

Thrombus Formation

Plaque disruption exposes highly thrombogenic subendothelial components of the vessel wall. Platelets adhere to these areas via specific receptor-mediated interactions. The most important appear to be the glycoprotein Ia/IIa interaction with collagen and the glycoprotein Ib/IX interaction with von Willebrand factor. As platelets are activated, they undergo a shape change, release granule constituents into the circulation, and rearrange their surface membrane with expression of the activated IIb/IIIa receptor and other integrins. This in turn stimulates circulating platelets and, via cross-linkage with fibrinogen, initiates aggregation. At the same time, the activated platelet membrane provides the milieu for the interaction of several of the clotting factors promoting the cleavage of fibrinogen and the formation of the fibrin network that stabilizes the growing platelet plug.

Detection of ongoing thrombosis in patients with suspected ACS via markers of both platelet activation and the fibrin pathway are under investigation. One such marker, P-selectin, is a member of the integrin family of receptors, which in the resting platelet is found within the α-granules. Upon platelet activation, P-selectin becomes exposed on the surface of the platelet and then is later cleaved off and can be detected free in the plasma. Assays for both membrane-bound and soluble P-selectin have been studied, but despite initial encouraging data, both have failed to provide the requisite level of sensitivity and specificity for accurate diagnosis of ACS patients, apparently due to a great amount of heterogeneity within the population.[28]

Markers that reflect activation of the coagulation cascade, including most of the conventional hemostatic markers, have also been widely studied. In general, these have not been able to distinguish patients having ACS with adequate sensitivity and specificity. However, a recent study by Li et al.[29] demonstrated that thrombin-antithrombin complex, fibrinopeptide A, and F1.2 were elevated in patients with AMI who died versus those who lived, suggesting that these assays may have prognostic value despite their poor diagnostic performance.

Reduced Blood Flow

The consequence of plaque rupture and platelet activation is a growing thrombus that may ultimately impair blood flow. The gold standard for defining reduced coronary blood flow is coronary angiography, ie, the TIMI flow criteria. Although angiography may not be appropriate for screening low-risk chest pain patients, technetium-based myocardial perfusion imaging (MPI) can noninvasively demonstrate impaired blood flow, either at rest or associated with provocative testing. We have used MPI extensively in apparently low-risk chest pain patients to identify those at high risk: the presence of perfusion defects, when associated with matched regional wall motion abnormalities, correlates with adverse clinical outcomes similar to those for patients with ECG abnormalities.[30,31]

Myocardial Ischemia

The mainstay for the detection of myocardial ischemia has been the ECG, where detection of the electrical criteria reflecting ischemic myocardium has good specificity. Unfortunately, the ECG has a number of limitations including a relatively low sensitivity; ST elevation is present in only 40% to 50% of patients who have an AMI,[32] and ECGs are often negative or nondiagnostic in patients with unstable angina.[3] Although an ischemic ECG can identify individuals who are at higher risk,[22,23] approximately 10% of AMI patients will have a normal initial ECG.[33,34] Several methods are being used to improve the sensitivity of the ECG. For instance, ST-segment trend monitoring can be used to identify some high risk-patients with a stuttering course who might be missed by a single tracing.[35,36] Extended-lead ECGs and body mapping are also being studied.[37] This is especially important for patients with ischemia in areas of the heart that tend to be electrographically quiescent on the standard 12-lead ECG, ie, the circumflex distribution to the posterior and lateral regions. Expanding the ECG to include posterior leads has been reported to improve sensitivity.[38,39]

Other methods to detect ischemia have also been investigated. Ischemic myocardium does not contract normally, so assessment of regional wall motion can be used to demonstrate ischemia. Gated MPI can provide information about regional wall motion and global left ventricular function. Real-time imaging with two-dimensional (2-D) echocardiography can also be used. Sabia and colleagues[40] demonstrated a significant increase in the adjusted cardiac event rate for patients with any wall motion abnormalities versus those with entirely normal left ventricular function. Kontos et al.[41] reported that 2D echocardiography is equivalent to MPI for identifying higher risk patients. However, both echocardiography and MPI require a high level of sophistication and logistic support to provide these services rapidly in the ED.

Ideally, a simple, rapid blood test to detect cardiac ischemia would be preferred, and several are now under intense scrutiny. Such assays should be able to detect all ACS patients in whom ischemia is ongoing or has recently occurred. Holvoet et al.[42] reported that malondialdehyde modified low-density lipoprotein (MDA-LDL) was elevated equally in patients with unstable angina and AMI, and thus could be used to identify higher risk individuals. In contrast, troponin I was significantly higher in AMI patients than in those with unstable angina.[42] An "Ischemia Test" based on ischemic modification of circulating serum albumin is reported to detect the recent occurrence of vascular ischemia.[43]

Markers of Myocardial Necrosis

The terminal event after prolonged ischemia is death of the myocardium and release of biochemical compounds reflective of necrosis into the circulation. Although much is made of the newer, more sensitive and specific markers, the fact remains that these are indicative of myocardial cell death, and thus any evaluation that relies on these assays as the sole determinant for the presence of an ACS may fail to identify a large proportion of the patients at risk.

The troponins, T and I (TnT and TnI), are demonstrating important new issues for the evaluation of patients with ACS. Both have cardiac-specific forms (cTnT, cTnI) that are ordinarily not detectable in serum. Their presence is thus indicative of myocardial damage, with the important distinction that they are detectable relatively early after infarction yet remain elevated for many days. These properties have important ramifications for the diagnosis of ACS. Katus et al.[44] demonstrated that the sensitivities of the MB isoenzyme of creatine kinase (CK-MB) and cTnT were similar among patients who were admitted for possible MI, though the specificity for cTnT was significantly lower than that for CK-MB despite its supposed absolute cardiac specificity. However, when patients with the clinical diagnosis of unstable angina were excluded from the analysis, the specificity for cTnT increased to the expected range.[44] This led to speculation that troponin could be used to diagnose unstable angina, and numerous published studies have examined the implications of elevations of troponin in the absence of CK-MB elevations.[45] Although not meeting the strict definition of MI, it is clear that these troponin elevations do result from myocardial damage and, irrespective of the clinical designation, carry prognostic information in the short as well as the long term.[45–49] Further, this has important clinical significance; in the CAPTURE study, non-ST-elevation ACS patients who were cTnT positive had a significant and substantial decrease in events after treatment with IIb/IIIa receptor blocker abciximab, whereas little benefit was seen in those patients who were cTnT negative.[50] Similar results have been reported for tirofiban and low molecular weight heparins.

Absence of troponin elevation does not necessarily indicate low risk. For example, in the paper by Galvani et al.,[45] the 30-day rate of MI or death in cTnI-negative Braunwald Class III unstable angina patients was still greater than 5%. The proportion of patients who have recurrent ischemia or who undergo revascularization remains high among those who are troponin negative, and is often not significantly different from those who are troponin positive.[47,51,52] That many patients have complications but do not have troponin elevations reflects the fact that sensitivity for all forms of ACS is suboptimal, and, therefore, patients without troponin elevations can be considered to be at lower risk but are not no-risk or even low-risk.[48,49] Although some would argue that any evidence of myocardial necrosis is indicative of AMI, other classifications have been suggested: the *Standards for Laboratory Practice*, published by the National Academy of Clinical Biochemistry, has endorsed the term "minor myocardial damage."[53]

Integration of Tests and Assays into Systematic Protocols

Biochemical markers, either alone or in combination, are inadequate to identify all patients with ACS. Although assays for markers of necrosis have improved dramatically, they will never detect a significant proportion of patients who have some earlier forms of ACS that do not progress to AMI. There are now tests and assays that can provide additional diagnostic or prognostic input into the ACS evaluation process (Fig. 2). The goal is to use this data in a fashion that maximizes the clinical outcomes. Within any population of patients presenting with chest pain, there is a combination of clinical factors that contributes

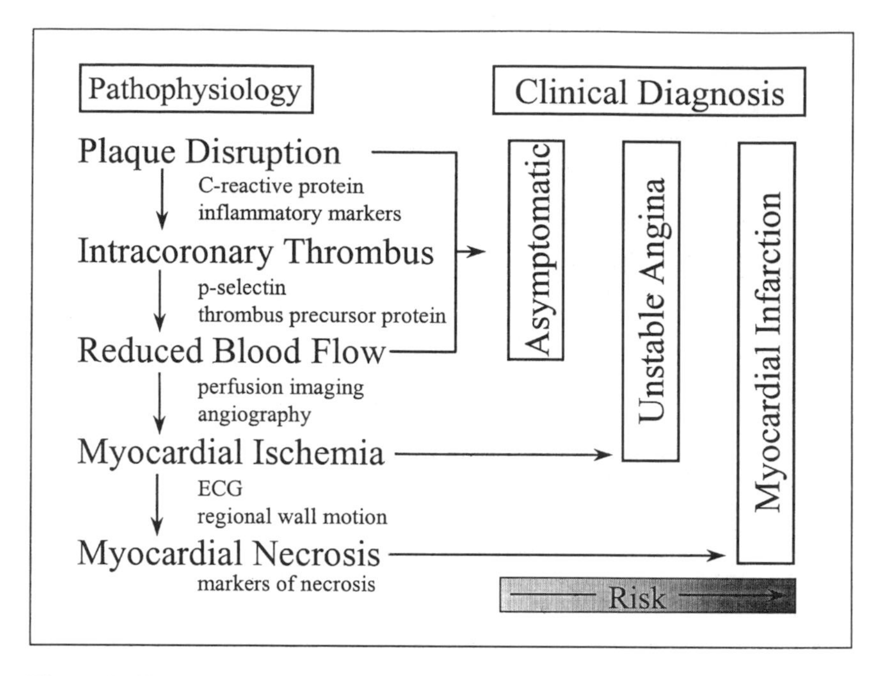

Figure 2. Tests and assays providing diagnostic and prognostic information in patients with an acute coronary syndrome. In the construct of the pathophysiological model for the acute coronary syndrome, there are either current or emerging opportunities for providing diagnostic prognostic information at each sequential step in the process.

to the risks and probability of an acute event. One approach is to analyze risk as a continuous variable and develop a population curve to fit this construct; the area under the curve represents the risk profile for an entire group of patients presenting with chest pain (Fig. 3A). For all intents and purposes, this was the practice taken in the ACI-TIPI study, in which the probability of acute ischemia was calculated via a combination of clinical and electrocardiographic inputs and reported as a continuum from 0% to 100%.[13] Although such algorithms can be accurate, it is often difficult to a apply them to direct patient care. For example, how can this array of probabilities be used in clinical decision making?

From a clinical perspective, it is more practical to assign risk in a manner that has clinical applicability. Since diagnostic (and prognostic) goals relate to the entirety of the ACS, the statements that define risk must include not only the probability of AMI but also the probability of acute ischemia. The Acute Cardiac Team (ACT) at the Medical College of Virginia Hospitals/Virginia Commonwealth University imple-

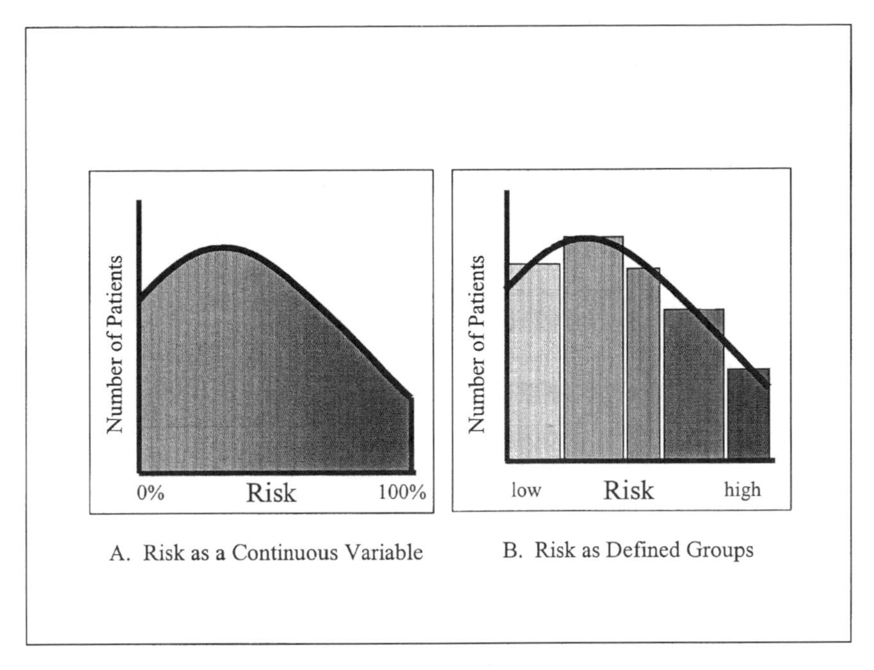

Figure 3. Distribution of risk among patients presenting with chest pain. A population model can be constructed that defines the risk within a given population of patients presenting with chest pain. Based on clinical findings and objective testing, a patient can be assigned to some level of risk. Panel A represents this risk as a continuous variable, whereas panel B groups risk into fewer discrete categories that can be associated with specific clinical categories defined by common treatment strategies or outcomes.

mented an integrated strategy for the evaluation of all patients presenting with chest pain. It was felt that a minimum of five levels of risk could define the chest pain population in a manner that supports clinical decision making (Fig. 3B). The fundamental premise was that the evaluation of chest pain must be approached through a hierarchical process that is 1) based on risk; 2) driven by specific goals for each level of risk; and 3) time-dependent (Fig. 4). For each risk level, a clinical guideline was developed that drives the initial care for patients assigned to that level.

Since June 1994, all patients presenting to the ED at the Medical College of Virginia Hospitals with chest pain or signs/symptoms suggestive of cardiac ischemia have undergone rapid evaluation with assignment to an ACT triage level (Fig. 5). The ACT level is assigned based on the probability of either AMI or myocardial ischemia, which is derived from clinical and ECG variables. Patients assigned to levels

Chest Pain Evaluation
A Hierarchy

Rapidly rule-in AMI and initiate therapy
↓
Rapidly rule-in unstable angina and initiate therapy
↓
Rule-out acute coronary syndromes
↓
Identify stable coronary disease
↓
Identify high risk individuals
↓
Risk factor modification

Figure 4. A hierarchy for the evaluation of the patient presenting with chest pain. The initial approach to all chest pain patients is defined by those at highest risk and the opportunity for risk reduction and outcome improvement. The focus does not stop with the rule-out of AMI, but must also include the rapid diagnosis and treatment of the entirety of the acute coronary syndrome. Secondarily, but no less importantly, is the identification of cardiovascular risk and the implementation of efforts to reduce that risk.

1, 2, and 3 are admitted, while patients in levels 4 and 5 are evaluated in the ED. Timing for the entire process is driven by the most urgent treatment modality, level 1 (AMI defined by ST-segment elevation), based on the well documented finding that revascularization with either fibrinolytics or primary angioplasty can improve mortality if delivered in a timely fashion. Thus, a fundamental concept for the protocol is that all patients must be considered to have an AMI until proven otherwise. It is important to note that biochemical markers of necrosis do not play a role for early risk stratification since assignment to a level and subsequent care are dictated by the ECG findings alone.

Level 2 also defines a high-risk group—those with unstable angina. ECG criteria include ST depression, ischemic T-wave inversions, or dynamic ST changes not meeting criteria for fibrinolytics. In addition, high-risk clinical findings, including signs and symptoms suggestive of impaired left ventricular function, are included among the entry

The ACT* Strategy for Evaluation and Triage of Chest Pain

Risk-Based, Goal-Driven, Time Dependent Strategy

Level	Risk	Goal		Time to Goal
		Primary	Secondary	
1	Very high	Intervention		30 min
2	High	Intervention	Diagnosis	30 min
3	Moderate	Diagnosis	Prevention	8 hours
4	Low	Prognosis	Prevention	3 hours
5	Very low	Alt Diag		N/A

Risk →

* ACT - the "Acute Cardiac Team" at MCV/VCU

Figure 5. The Acute Cardiac Team (ACT) protocol for the evaluation and treatment of patients presenting with chest pain. The ACT at the Medical College of Virginia Campus at Virginia Commonwealth University uses a systematic protocol for the initial triage and treatment of chest pain. This is composed of risk-based, goal-driven, and time-dependent pathways that direct the early management of this patient population.

criteria. Here, biochemical markers of necrosis are imperative, as a significant percentage of these patients will in fact rule in for an infarction. The integration of a marker strategy into this pathway depends on the subsequent treatments, eg, early administration of IIb/IIIa receptor blockers or immediate catheterization driven by biochemical markers would mandate a strategy constructed for early detection of the positive patients.

The criteria for levels 3 and 4 are less definitive. Since diagnostic ECG findings are lacking, the integration of additional technology is required. Using the construct from Figure 2, it is obvious that the yield of biochemical markers becomes lower as the relative risk of AMI decreases, and the diagnostic strategy must shift toward the detection of unstable angina. Here, the need is for technology aimed at detection of ischemia, reduced blood flow, thrombosis, or plaque rupture. While there are promising assays in early stages of clinical development, at present there are no biochemical markers available for any of these.

Imaging and provocative testing have thus taken a pivotal role in the triage of the lower risk chest pain patient.

The ACT protocol uses acute rest MPI for the identification of unstable angina because it can provide information about both reduced blood flow (myocardial perfusion) and myocardial ischemia (regional wall motion). However, the clinical goal of acute perfusion imaging is different in level 3 (diagnostic) versus level 4 (prognostic). In levels 3 and 4, follow-up stress testing is used to exclude significant coronary disease. Acute MPI can predict outcomes in chest pain patients initially considered at low to intermediate risk by identifying individuals who are at higher risk than suggested by the history, physical examination, and ECG. When one compares cardiovascular events in patients with positive versus negative MPI, there is a significant increase in MI and revascularization, and significant disease is present in patients with a positive MIBI stress test (Fig. 6).[54]

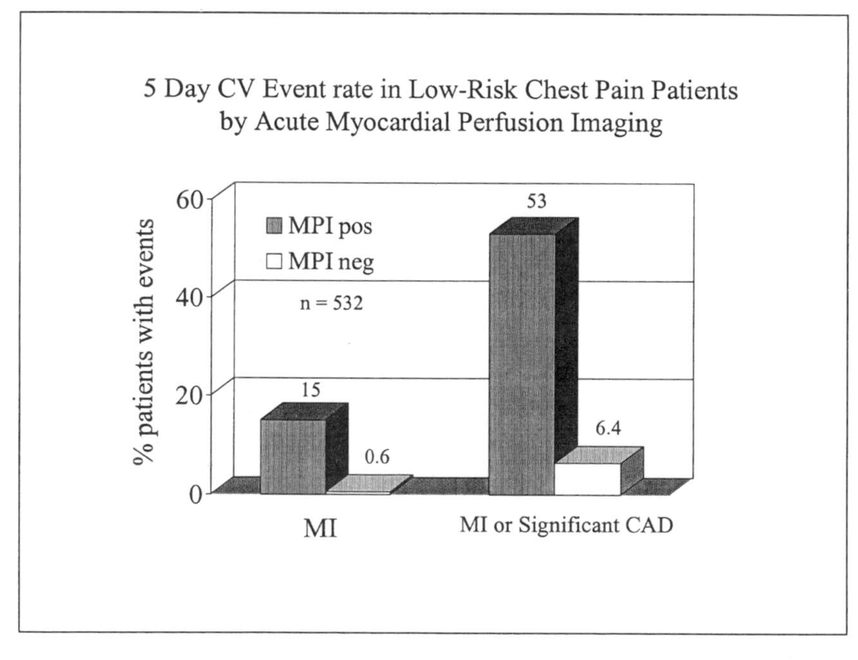

Figure 6. Cardiovascular outcomes for low-risk chest pain patients based on the results of rest sestamibi myocardial perfusion imaging. The use of perfusion imaging to identify high-risk patients among those who appear to be at lower risk based on clinical and electrocardiographic findings is evident for both acute myocardial infarction and the finding of significant coronary disease. Significant disease is defined as angiographic lesions greater than 70% or the occurrence of revascularization.

Acute MPI does have limitations, including the inability to distinguish acute infarction, acute ischemia, or old infarction, all of which are associated with perfusion defects. Further, the sensitivity of MPI decreases as the symptom-free period increases, although several studies have found a high sensitivity despite the absence of symptoms at the time of injection.[30,55–57] In our analysis of moderate- to low-risk ACT patients, the sensitivity of MPI was significantly higher than that of TnI for predicting non-AMI cardiac outcomes, and MPI was able to identify the higher risk patients more rapidly than was TnI.[58] It is important to note that the AMI rate is predictively low among moderate- to low-risk patients although the ACS rate is considerably higher: while only 4% of level 3 patients had AMI, 13% had AMI or revascularization, which is consistent with the ACS risk model discussed above. Further, the AMI rate for level 4 was even lower, only 0.9%. However, level 4 comprised nearly half of all patients, so this low relative percentage represented 6% of the total number of AMIs, a proportion similar to the published rate of missed AMIs.

The ACT experience with MPI is an excellent validation for the integration of diagnostic/prognostic testing into a risk-based systematic pathway. Other protocols, such as those used in chest pain evaluation units at the University of Cincinnati, at the Mayo Clinic, and at the University of California at Davis, to name but a few, have integrated other technologies with similar results.[36,59,60] One important concept common to all is that risk can be defined by a finite number of categories, which in turn can describe specific triage/diagnostic/treatment protocols.

A second important concept is that the testing implemented for each risk level must be able to meet the defined goal of that level, and must ultimately be able to influence outcomes. Any given test may also be used for different reasons at various levels of risk. For instance, biochemical markers of necrosis may determine early therapeutic options in unstable angina, whereas in AMI, biochemical markers are used primarily for late risk stratification. In a similar fashion, the role for acute imaging is different for intermediate-risk patients, in whom it is used to rule in ACS, versus low-risk patients, for whom it is a prognostic test to assure safe discharge until the definitive stress test can be performed.

Conclusion

The effective evaluation of chest pain patients is dependent on risk stratification driven initially by the history, physical examination, and ECG. However, these frequently fail to reach the level of sensitivity needed to prevent inadvertent discharge of a small but significant proportion of higher risk individuals with occult symptoms, and thus further objective testing must be integrated into the evaluation process. Biochemical markers of necrosis are often not useful in the lower risk patients because they lack sensitivity for ischemia. For this simple reason, we are forced to consider other technologies, and, fortunately, the ability to detect changes in myocardial blood flow or to analyze other surrogates of ischemia does appear to have clinical value. Several risk stratification models have been studied; acute stress testing, MPI, and 2-D echo appear efficacious but suffer major logistical limitations to widespread application. Further, no one technology can both meet the needs of all patients and be implemented in all medical centers. Thus, the choice of technologies must be based on local expertise and resources.

Fortunately, biochemical assays that are capable of identifying patients with vascular ischemia are beginning to emerge and are now in the early stages of clinical testing. If validated, these assays would facilitate the identification of higher risk patients among those who appear to be at low risk based on initial clinical assessment. Use of these assays will be dependent upon their ability to provide an acceptable level of sensitivity and specificity. If this is indeed the case, rapid and widespread implementation is likely, as these would not be subject to the logistic limitations that encumber most imaging modalities. As such, when integrated into risk-based systematic protocols, these assays would have the potential to further reduce cost and improve patient outcomes.

References

1. Gibler WB. Chest pain units: Do they make sense now? *Ann Emerg Med* 1997;29:168–171.
2. Lee TH, Cook EF, Weisberg M, et al. Acute chest pain in the emergency room. Identification and examination of low-risk patients. *Arch Intern Med* 1985;145:65–69.
3. Lee TH, Rouan GW, Weisberg MC, et al. Clinical characteristics and natural history of patients with acute myocardial infarction sent home from the emergency room. *Am J Cardiol* 1987;60:219–224.
4. McCarthy BD, Beshansky JR, D'Agostino RB, Selker HP. Missed diagnoses of acute myocardial infarction in the emergency department: Results from a multicenter study. *Ann Emerg Med* 1993;22:579–582.
5. Tierney WM, Fitzgerald J, McHenry R, et al. Physicians' estimates of the probability of myocardial infarction in emergency room patients with chest pain. *Med Decis Making* 1986;6:12–17.
6. Rouan GW, Hedges JR, Toltzis R, et al. A chest pain clinic to improve the follow-up of patients released from an urban university teaching hospital emergency department. *Ann Emerg Med* 1987;16:1145–1150.

7. Goldman L, Cook EF, Brand DA, et al. A computer protocol to predict myocardial infarction in emergency department patients with chest pain. *N Engl J Med* 1988;318:797–803.
8. Puleo PR, Meyer D, Wathen C, et al. Use of a rapid assay of subforms of creatine kinase-MB to diagnose or rule out acute myocardial infarction. *N Engl J Med* 1994;331:561–566.
9. Selker HP, Griffith JL, Dorey FJ, D'Agostino RB. How do physicians adapt when the coronary care unit is full? A prospective multicenter study. *JAMA* 1987;257:1181–1185.
10. Graff LG, Dallara J, Ross MA, et al. Impact on the care of the emergency department chest pain patient from the chest pain evaluation registry (CHEPER) study. *Am J Cardiol* 1997;80:563–568.
11. Ting HH, Lee TH, Soukup JR, et al. Impact of physician experience on triage of emergency room patients with acute chest pain at three teaching hospitals. *Am J Med* 1991;91:401–408.
12. Pozen MW, D'Agostino RB, Mitchell JB, et al. The usefulness of a predictive instrument to reduce inappropriate admissions to the coronary care unit. *Ann Intern Med* 1980;92:238–242.
13. Selker HP, Beshansky JR, Griffith JL, et al. Use of the acute cardiac ischemia time-insensitive predictive instrument (ACI-TIPI) to assist with triage of patients with chest pain or other symptoms suggestive of acute cardiac ischemia. A multicenter, controlled clinical trial. *Ann Intern Med* 1998;129: 845–855.
14. Nomenclature and criteria for diagnosis of ischemic heart disease. Report of the Joint International Society and Federation of Cardiology/World Health Organization task force on standardization of clinical nomenclature. *Circulation* 1979;59:607–609.
15. van Miltenburg-van Zijl AJ, Simoons ML, Veerhoek RJ, Bossuyt PM. Incidence and follow-up of Braunwald subgroups in unstable angina pectoris. *J Am Coll Cardiol* 1995;25:1286–1292.
16. Calvin JE, Klein LW, VandenBerg BJ, et al. Risk stratification in unstable angina. Prospective validation of the Braunwald classification. *JAMA* 1995; 273:136–141.
17. Selker HP, Griffith JL, D'Agostino RB. A tool for judging coronary care unit admission appropriateness, valid for both real-time and retrospective use. A time-insensitive predictive instrument (TIPI) for acute cardiac ischemia: A multicenter study. *Med Care* 1991;29:610–627.
18. Goldman L, Cook EF, Johnson PA, et al. Prediction of the need for intensive care in patients who come to the emergency departments with acute chest pain. *N Engl J Med* 1996;334:1498–1504.
19. Killip T III, Kimball JT. Treatment of myocardial infarction in a coronary care unit. A two year experience with 250 patients. *Am J Cardiol* 1967;20: 457–464.
20. Chin MH, Cook EF, Lee TH, Goldman L. Correlates of major complications and mortality in patients presenting to the emergency department with chest pain and more than bibasilar rales. *J Gen Intern Med* 1994;9:659–665.
21. Timmis AD. Early diagnosis of acute myocardial infarction. *Br Med J* 1990; 301:941–942.
22. Savonitto S, Ardissino D, Granger CB, et al. Prognostic value of the admission electrocardiogram in acute coronary syndromes. *JAMA* 1999;281: 707–713.
23. Holmvang L, Clemmensen P, Wagner G, Grande P. Admission standard

electrocardiogram for early risk stratification in patients with unstable coronary artery disease not eligible for acute revascularization therapy: A TRIM substudy. Thrombin Inhibition in Myocardial Infarction. *Am Heart J* 1999; 137:24–33.

24. Zhu J, Quyyumi AA, Norman JE, et al. Cytomegalovirus in the pathogenesis of atherosclerosis: The role of inflammation as reflected by elevated C-reactive protein levels. *J Am Coll Cardiol* 1999;34:1738–1743.
25. Liuzzo G, Biasucci LM, Gallimore JR, et al. The prognostic value of C-reactive protein and serum amyloid a protein in severe unstable angina. *N Engl J Med* 1994;331:417–424.
26. Abdelmouttaleb I, Danchin N, Ilardo C, et al. C-Reactive protein and coronary artery disease: Additional evidence of the implication of an inflammatory process in acute coronary syndromes. *Am Heart J* 1999;137:346–351.
27. Ridker PM, Cushman M, Stampfer MJ, et al. Inflammation, aspirin, and the risk of cardiovascular disease in apparently healthy men. *N Engl J Med* 1997;336:973–979.
28. Hollander JE, Muttreja MR, Dalesandro MR, Shofer FS. Risk stratification of emergency department patients with acute coronary syndromes using P-selectin. *J Am Coll Cardiol* 1999;34:95–105.
29. Li YH, Teng JK, Tsai WC, et al. Prognostic significance of elevated hemostatic markers in patients with acute myocardial infarction. *J Am Coll Cardiol* 1999;33:1543–1548.
30. Kontos MC, Jesse RL, Schmidt KL, et al. Value of acute rest sestamibi perfusion imaging for evaluation of patients admitted to the emergency department with chest pain. *J Am Coll Cardiol* 1997;30:976–982.
31. Tatum JL, Jesse RL, Kontos MC, et al. Comprehensive strategy for the evaluation and triage of the chest pain patient. *Ann Emerg Med* 1997;29: 116–123.
32. Rouan GW, Lee TH, Cook EF, et al. Clinical characteristics and outcome of acute myocardial infarction in patients with initially normal or nonspecific electrocardiograms (a report from the Multicenter Chest Pain Study). *Am J Cardiol* 1989;64:1087–1092.
33. Slater DK, Hlatky MA, Mark DB, et al. Outcome in suspected acute myocardial infarction with normal or minimally abnormal admission electrocardiographic findings. *Am J Cardiol* 1987;60:766–770.
34. Brush JE Jr, Brand DA, Acampora D, et al. Use of the initial electrocardiogram to predict in-hospital complications of acute myocardial infarction. *N Engl J Med* 1985;312:1137–1141.
35. Fesmire FM, Percy RF, Bardoner JB, et al. Usefulness of automated serial 12-lead ECG monitoring during the initial emergency department evaluation of patients with chest pain. *Ann Emerg Med* 1998;31:3–11.
36. Gibler WB, Runyon JP, Levy RC, et al. A rapid diagnostic and treatment center for patients with chest pain in the emergency department. *Ann Emerg Med* 1995;25:1–8.
37. Kornreich F, Montague TJ, Rautaharju PM. Body surface potential mapping of ST segment changes in acute myocardial infarction. Implications for ECG enrollment criteria for thrombolytic therapy. *Circulation* 1993;87:773–782.
38. Zalenski RJ, Rydman RJ, Sloan EP, et al. Value of posterior and right ventricular leads in comparison to the standard 12-lead electrocardiogram in evaluation of ST-segment elevation in suspected acute myocardial infarction. *Am J Cardiol* 1997;79:1579–1585.
39. Zalenski RJ, Cooke D, Rydman R, et al. Assessing the diagnostic value of

an ECG containing leads V4R, V8, and V9: The 15-lead ECG. *Ann Emerg Med* 1993;22:786–793.

40. Sabia P, Abbott RD, Afrookteh A, et al. Importance of two-dimensional echocardiographic assessment of left ventricular systolic function in patients presenting to the emergency room with cardiac-related symptoms. *Circulation* 1991;84:1615–1624.
41. Kontos MC, Arrowood JA, Jesse RL, et al. Comparison of echocardiography and myocardial perfusion imaging for diagnosing emergency department patients with chest pain. *Am Heart J* 1998;136:724–733.
42. Holvoet P, Collen D, Van de Werf F. Malondialdehyde-modified LDL as a marker of acute coronary syndromes. *JAMA* 1999;281:1718–1721.
43. Bar-Or D, Lau E, Rao N, et al. Reduction in the cobalt binding capacity of human albumin with myocardial ischemia. *Ann Emerg Med* 1999;34:4
44. Katus HA, Remppis A, Neumann FJ, et al. Diagnostic efficiency of troponin T measurements in acute myocardial infarction. *Circulation* 1991;83: 902–912.
45. Galvani M, Ottani F, Ferrini D, et al. Prognostic influence of elevated values of cardiac troponin I in patients with unstable angina. *Circulation* 1997;95: 2053–2059.
46. Lindahl B, Venge P, Wallentin L. Relation between troponin T and the risk of subsequent cardiac events in unstable coronary artery disease. The FRISC study group. *Circulation* 1996;93:1651–1657.
47. Antman EM, Tanasijevic MJ, Thompson B, et al. Cardiac-specific troponin I levels to predict the risk of mortality in patients with acute coronary syndromes. *N Engl J Med* 1996;335:1342–1349.
48. Polanczyk CA, Lee TH, Cook EF, et al. Cardiac troponin I as a predictor of major cardiac events in emergency department patients with acute chest pain. *J Am Coll Cardiol* 1998;32:8–14.
49. Sayre MR, Kaufmann KH, Chen IW, et al. Measurement of cardiac troponin T is an effective method for predicting complications among emergency department patients with chest pain. *Ann Emerg Med* 1998;31:539–549.
50. Hamm CW, Heeschen C, Goldmann B, et al. Benefit of abciximab in patients with refractory unstable angina in relation to serum troponin T levels. c7E3 ab Antiplatelet Therapy in Unstable Refractory Angina (CAPTURE) Study Investigators. *N Engl J Med* 1999;340:1623–1629.
51. Luscher MS, Thygesen K, Ravkilde J, Heickendorff L. Applicability of cardiac troponin T and I for early risk stratification in unstable coronary artery disease. TRIM Study Group. Thrombin Inhibition in Myocardial ischemia. *Circulation* 1997;96:2578–2585.
52. Benamer H, Steg PG, Benessiano J, et al. Elevated cardiac troponin I predicts a high-risk angiographic anatomy of the culprit lesion in unstable angina. *Am Heart J* 1999;137:815–820.
53. Wu AHB, Apple FS, Warshaw MM, et al. National Academy of Clinical Biochemistry Standards of Laboratory Practice recommendations for use of cardiac markers in coronary artery disease. *Clin Chem* 1999;45:1104–1121.
54. Kontos MC, Jesse RL, Schmidt KL, et al. Value of acute rest sestamibi perfusion imaging for evaluation of patients admitted to the emergency department with chest pain. *J Am Coll Cardiol* 1997;30:976–982.
55. Wackers FJ, Lie KI, Liem KL, et al. Potential value of thallium-201 scintigraphy as a means of selecting patients for the coronary care unit. *Br Heart J* 1979;41:111–117.
56. Varetto T, Cantalupi D, Altieri A, Orlandi C. Emergency room technetium-

99m sestamibi imaging to rule out acute myocardial ischemic events in patients with nondiagnostic electrocardiograms. *J Am Coll Cardiol* 1993;22: 1804–1808.
57. Bilodeau L, Theroux P, Gregoire J, et al. Technetium-99m sestamibi tomography in patients with spontaneous chest pain: Correlations with clinical, electrocardiographic and angiographic findings [see comments]. *J Am Coll Cardiol* 1991;18:1684–1691.
58. Kontos MC, Jesse RL, Anderson FP, et al. Comparison of myocardial perfusion imaging and cardiac troponin I in patients admitted to the emergency department with chest pain. *Circulation* 1999;99:2073–2078.
59. Kirk JD, Turnipseed S, Lewis WR, Amsterdam EA. Evaluation of chest pain in low-risk patients presenting to the emergency department: The role of immediate exercise testing. *Ann Emerg Med* 1998;32:1–7.
60. Farkouh ME, Smars PA, Reeder GS, et al, for the Chest Pain Evaluation in the Emergency Room (CHEER) Investigators. A clinical trial of a chest-pain observation unit for patients with unstable angina. *N Engl J Med* 1998; 339:1882–1888.

Chapter 9

The Use of Cardiac Markers for Therapeutic Decisions in Acute Coronary Syndromes

Michael P. Hudson, MD, Britta U. Goldmann, MD, and E. Magnus Ohman, MD

Introduction

The acute coronary syndromes (ACS) include the conditions of ST-segment elevation myocardial infarction (MI), unstable angina, and non-ST-segment elevation MI. These conditions share a pathologic mechanism of atherosclerotic plaque disruption and lead to coronary artery platelet aggregation and thrombosis. Myocardial ischemia or necrosis then results, as indicated by raised concentrations of cardiac markers in the bloodstream. Certain markers present in cardiac muscle, such as creatine kinase (CK), CK-MB isoenzyme, and cardiac-specific troponins (T and I), are released into the peripheral circulation when myocardial necrosis occurs. The timing and concentration of these macromolecules in the circulation depend on local blood flow, infarct related artery patency, infarct size, and each marker's intracellular location, molecular weight, and elimination rate from the blood.

Over the last two decades, physicians have increasingly relied on more cardiac-specific markers (CK-MB, cardiac troponin T and I [cTnT, cTnI]) to detect myocardial necrosis. The implementation of these newer cardiac markers, particularly the cardiac troponins, greatly improves the diagnosis and risk stratification of patients with ACS. Both cTnT and cTnI are present in high concentrations in the myocardium and are absent in nonmyocardial tissue and serum, thus preventing false-positive elevations due to skeletal muscle, renal, or gastrointestinal injury. Sensitivity to detect small quantities of myocardial necrosis

From: Adams JE III, Apple FS, Jaffe AS, Wu AHB (eds). *Markers in Cardiology: Current and Future Clinical Applications.* Armonk, NY: Futura Publishing Company, Inc.; © 2001.

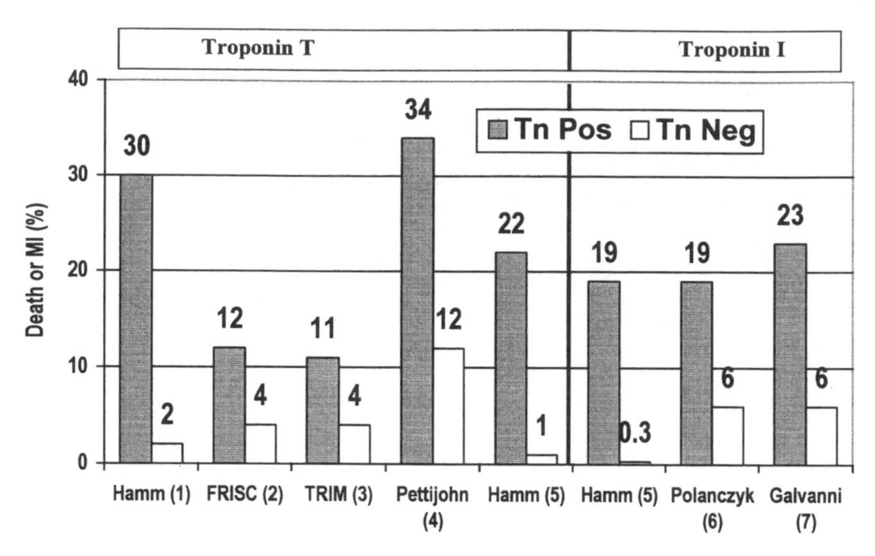

Figure 1. Prognostic value of troponin T and troponin I on the incidence of death or myocardial infarction after hospital presentation for acute coronary syndromes.

is also enhanced. The cardiac troponins detect myocardial injury in 30% to 40% of patients with myocardial ischemia and normal values of CK and CK-MB.[1,2]

Elevated concentrations of cTnT and cTnI contribute prognostic data beyond electrocardiogram and CK-MB results. ACS trials have consistently shown that raised levels of cTnT or cTnI predict threefold or greater risk of death or death/MI, and that cardiac troponin values are independently associated with increased mortality in ACS patients (Fig. 1).[1–7] In the TIMI IIIB trial, 6-week mortality was significantly higher in patients with cTnI levels of 0.4 μg/L or higher than in patients with cTnI levels below 0.4 μg/L (3.7% versus 1.0%; $P<0.001$).[8] Similarly in the GUSTO IIa trial, 30-day mortality was significantly higher in patients with cTnT levels greater than 0.1 μg/L than in patients with lower levels of cTnT (11.8% versus 3.9%; $P<0.001$).[9] Measurements of troponin levels may be of greatest value in patients with normal CK/CK-MB levels. In such patients, the presence of an elevated cTnT value is associated with an odds ratio of 3.9 for coronary events within the following 6 months.[4]

A New Cardiac Marker Paradigm

Traditionally, biochemical markers of myocardial necrosis have been collected serially over 6 to 48 hours in ACS patients to confirm or exclude the diagnosis of acute MI (AMI) and to estimate infarct size. World Health Organization diagnostic criteria specify an elevation and subsequent fall of cardiac marker(s) as proof of AMI.[10] Similarly, physicians use serial cardiac marker results to differentiate unstable angina from non-ST-segment elevation MI in patients with similar clinical presentations. These applications rely on cardiac marker results available hours after patient presentation, and fail to couple cardiac marker data with time-dependent and evidence-based early therapies.

In a new paradigm, cardiac marker testing may assist physicians to better target more potent antithrombotic therapies for higher risk patients. Thirty-six percent of ST-segment elevation MI deaths and 12% of non-ST-segment elevation ACS deaths occur within 24 hours of hospital presentation.[11] Time-dependent therapies and the pace of ACS decision making will not allow for prolonged assay times and serial laboratory results. Just as electrocardiographic ST-segment elevation identifies patients who benefit from immediate reperfusion therapy, elevated concentrations of cardiac markers may identify ACS patients who benefit from more potent antiplatelet, antithrombin, and interventional approaches. Use of such an early, reliable risk-stratification process will optimize clinical outcomes and the economical allocation of medical resources.

Therapy Implications

Proper risk stratification and use of cardiac marker results may lead to improved treatment decisions. Recent clinical trial analyses with prospectively collected cardiac markers demonstrate that glycoprotein IIb/IIIa (GP IIb/IIIa) inhibition and low molecular weight heparin administration exert clinical benefits predominantly in patients with elevated levels of cardiac markers.[12–14] These studies are summarized below.

GP IIb/IIIa Inhibition (CAPTURE Trial)

The platelet GP IIb/IIIa inhibitors significantly reduce nonfatal MI and the need for repeat procedures in patients undergoing percutaneous coronary intervention or presenting with an ACS.[15] Widespread application of these agents has been hampered by questions of safety, modest treatment effect, and cost. Cardiac marker results may assist physicians to better target these agents for patients at higher risk.

The c7E3 Fab Antiplatelet Therapy in Unstable Refractory Angina (CAPTURE) trial demonstrated that treatment with the GP IIb/IIIa antagonist abciximab reduced the risk of MI in patients with refractory unstable angina, during the 18- to 24-hour period preceding the coronary intervention and during the subsequent balloon angioplasty.[16] Hamm et al.[12] subsequently reported that 31% of CAPTURE patients had elevated levels (>0.1 μg/L) of cTnT at study entry. The 6-month cumulative rate of death or nonfatal MI was 16.7% among patients with elevated cTnT levels, as compared with 8.5% among patients without elevated cTnT levels ($P \leq 0.001$). Among patients with elevated cTnT levels, abciximab significantly reduced 6-month event rates versus placebo (9.5% versus 23.9%, odds ratio 0.32 [95% CI 0.14 to 0.62]; $P=0.002$) (Fig. 2). Conversely, in patients without elevated cTnT levels, there was no benefit of abciximab versus placebo in reducing 6-month death or MI (9.4% versus 7.5%, odds ratio 1.26 [95% CI 0.74 to 2.31]; $P=0.47$). Thus, abciximab significantly reduced the 6-month risk of death or MI versus placebo only in patients with elevated levels of cTnT.

An angiographic substudy of the CAPTURE study provides a mechanistic link between cTnT elevation and abciximab treatment efficacy. Baseline angiographic and cTnT data were correlated with subsequent angiographic findings obtained after 18 to 24 hours of abciximab versus placebo administration. Heeschen et al.[17] reported that complex lesion characteristics, lower TIMI flow rates, and the presence of visible thrombi were significantly linked to baseline cTnT elevation in CAP-

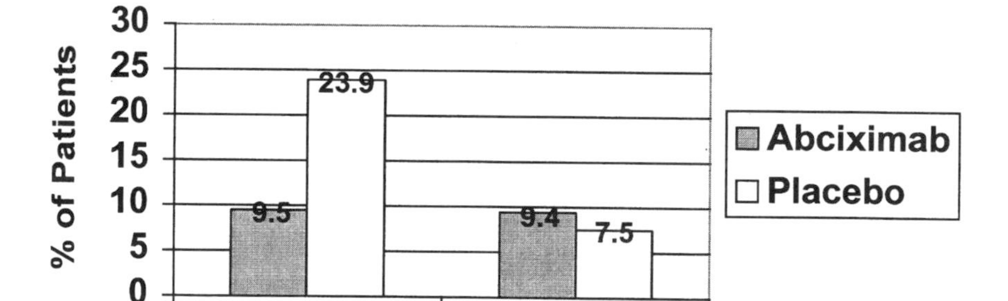

Figure 2. CAPTURE Trial. Six-month incidence of death or nonfatal myocardial infarction (MI) based on serum troponin T drawn at study entry. (Positive troponin T: >0.1 μg/L.)

TURE patients. In cTnT-negative patients, neither abciximab nor placebo resulted in improved TIMI flow rates on repeat angiography. In cTnT-positive patients, abciximab reduced the rate of TIMI flow ≤ 1 (16.1% → 12.5%, relative reduction = 22.4%) as compared with placebo (14.9% → 17.1%, relative increase = 14.8%) ($P \leq 0.02$). Similarly, abciximab produced greater frequency of thrombus resolution compared with placebo in cTnT-positive patients (63% versus 24%; $P \leq 0.001$), whereas no significant reduction of thrombus was seen in cTnT-negative patients for either abciximab (1%) or placebo (1%).

GP IIb/IIIa Inhibition (PRISM Trial, PRISM-PLUS Trial)

In the Platelet Receptor Inhibition in Ischemic Syndrome Management (PRISM) trial, a significant reduction in the rate of death, MI, and refractory ischemia was seen during 48-hour tirofiban infusion compared with unfractionated heparin.[18] At 30 days, there was no significant effect on the combined endpoint, thus prompting further investigation into identifying those ACS patients most likely to benefit from GP IIb/IIIa inhibition. Heeschen et al.[13] therefore investigated whether cTnT or cTnI could reliably identify high-risk patients who would benefit preferentially from tirofiban.

Of 3232 patients from the PRISM study, 2222 had baseline measurements of either cTnT or cTnI. Six hundred and twenty-nine patients (28.3%) had cTnI concentrations higher than the diagnostic threshold of 1.0 μg/L, and 644 patients (29.0%) had cTnT concentrations higher than 0.1 μg/L. Thirty-day event rates (death or MI) were 13.0% for cTnI-positive patients compared with 4.9% for cTnI-negative patients ($P \leq 0.001$), and 13.7% for cTnT-positive patients versus 3.5% for cTnT-negative patients ($P \leq 0.001$). At 30 days, among cTnI-positive patients, tirofiban lowered the incidence of death (adjusted hazard ratio 0.25 [95% CI 0.09 to 0.68]; $P = 0.004$) and MI (0.37 [95% CI 0.16 to 0.84]; $P = 0.01$). No treatment effect of tirofiban over unfractionated heparin was seen for cTnI-negative patients (Table 1). Similar effects with tirofiban were seen for cTnT-positive versus cTnT-negative patients.[13]

In the PRISM-PLUS trial, treatment with tirofiban plus heparin significantly reduced the 7-day incidence of death, MI, or refractory ischemia versus heparin therapy alone.[19] Among a small cohort of PRISM-PLUS patients, Hahn et al.[20] investigated the effect of tirofiban on early cTnI levels. Baseline cTnI levels did not differ between patients randomized to heparin (3.1 ± 6.7 μg/L) versus combination therapy (1.6 ± 3.0 μg/L). However, the peak 24-hour cTnI level was significantly higher in the heparin group (15.5 ± 29.1 μg/L) as compared with the

Table 1

PRISM Study—30-Day Event Rates for Refractory Ischemia, Death/MI, Death, MI (by Troponin I Status)

	Troponin I Positive		Troponin I Negative	
30-Day Event Rates	Heparin (n=324)	Tirofiban (n=305)	Heparin (n=801)	Tirofiban (n=792)
Death/MI	42 (13.0%)	13 (4.3%)*	39 (4.9%)	45 (5.7%)†
Death	20 (6.2%)	5 (1.6%)	18 (2.3%)	18 (2.3%)
MI	22 (6.8%)	8 (2.6%)	21 (2.6%)	27 (3.6%)
Refractory Ischemia	48 (14.8%)	31 (10.2%)	60 (7.5%)	72 (9.1%)

*Adjusted Hazard Ratio: 0.25 [95% CI 0.09–0.84], $P<0.001$; †Adjusted Hazard Ratio: 1.17 [95% CI 0.76–1.83], $P = 0.50$. PRISM = Platelet Receptor Inhibition in Ischemic Syndrome Management; MI = myocardial infarction.

combination therapy group (5.2 ± 8.3 μg/L) ($P=0.017$). The authors concluded that this prevention of early myocardial injury and necrosis might be responsible for later reduction of adverse cardiac events.

Long-Term Antithrombotic Therapy (FRISC Trial)

Antithrombin therapy in addition to aspirin antiplatelet therapy reduces event rates in patients with unstable angina and non-ST-elevation MI.[21] In the Fragmin in Unstable Coronary Artery Disease (FRISC) study, low molecular weight heparin (dalteparin) together with aspirin was more effective than aspirin alone in reducing 6-day mortality (1.8% versus 3.8%; $P<0.001$).[22] In this trial, all patients received subcutaneous dalteparin or placebo twice daily for 6 days with once daily dalteparin or placebo continued for another 5 weeks. Extending dalteparin treatment to 6 weeks and measuring outcomes at 6 months, however, produced nonsignificant treatment effects; therefore, Lindahl et al.[14] investigated prospectively whether different levels of peak cTnT might identify ACS patients likely to benefit from short- or long-term therapy with dalteparin.

At 6 days, dalteparin reduced the incidence of death or MI from 2.4% to 0% (n=327, $P=0.12$) and from 6.0% to 2.5% (n=644, $P<0.05$) in patients with cTnT levels of less than 0.1 and $\geq$0.1 μg/L, respectively.[14] During long-term treatment, a significant treatment effect of dalteparin versus placebo emerged only in cTnT-positive patients. In patients with cTnT $\geq$0.1 μg/L, dalteparin reduced the 40-day incidence of death or MI (7.4% versus 14.2%, relative risk = 0.52 [95% CI 0.32 to 0.83];

Table 2

FRISC Trial—Effect of Dalteparin on the Rate of Death or Myocardial Infarction (day 40) in Relation to Troponin T Level (<0.1; or ≥0.1 μg/L at Inclusion (n=971)

Troponin T (Placebo/Dalteparin)	Death or MI, Day 40 Placebo [no. (%)]	Dalteparin [no. (%)]	RR (95% CI)	*P*
<0.1 μg/L (170/157)	8 (4.7)	9 (5.7)	1.22 (0.48–3.1)	0.68
≥0.1 μg/L (318/326)	45 (14.2)	24 (7.4)	0.52 (0.32–0.83)	<0.01

FRISC = Fragmin in Unstable Coronary Artery Disease; CI = confidence interval; MI = myocardial infarction; RR = relative risk.

P<0.01), whereas no beneficial effect was seen in patients with cTnT less than 0.1 μg/L (5.7% versus 4.7%, relative risk =1.22 [0.48 to 3.1]; *P*-ns) (Table 2). Thus, elevation of cTnT may identify a subgroup of ACS patients in whom prolonged antithrombotic therapies is beneficial.

Future Cardiac Marker Applications

New cardiac markers, new applications of existing markers, and increasingly rapid and more convenient point-of-care assays are being developed. New Zealand investigators have demonstrated that patients with ST-segment elevation MI presenting with cTnT ≥0.1 μg/L are less likely to achieve TIMI grade 3 flow after intravenous thrombolysis (32% versus 62% [cTnT<0.1 μg/L]; *P*=0.02).[23] We have investigated the prognostic ability of a "point of care" qualitative cTnT assay (Cardiac T/TROP T, Boehringer Mannheim [now Roche, Indianapolis, IN]) to predict 90-minute reperfusion success and survival in ST-segment elevation MI patients.[24] In this analysis, 9.3% of ST-elevation MI patients had positive qualitative cTnT results at hospital presentation (184/1985). Positive cTnT results at hospital presentation were associated with lower rates of complete (>70%) ST-segment resolution (31% versus 46%; *P*<0.01) and higher rates of no (<30%) ST-segment resolution (37% versus 26%; *P*<0.01) after thrombolysis (Table 3). Patients with elevated baseline cTnT results had significantly higher 30-day mortality (15.2% versus 5.3% for cTnT-negative patients; *P*=0.001). A positive cTnT result added independently to the prediction of 30-day

Table 3

GUSTO III Qualitative Cardiac Troponin T and ST-Segment Resolution Substudy—Association of Cardiac Troponin T Result with Categories of ST-Segment Resolution and 30-Day Mortality

90-minute ST Resolution	cTnT Negative (n=1801)	cTnT Positive (n=185)	*P*
None, <30%	25.8% (464)	36.8% (68)	0.001
Partial, 30–70%	28.4% (512)	31.9% (59)	
Complete, ≥70%	45.8% (825)	30.8% (57)	
30-day Mortality	5.3%	15.2%	0.001

mortality in a multivariable model including age, blood pressure, heart rate, Killip class, MI location, and ST-segment resolution (χ^2 = 4.2; $P = 0.04$).

Conclusion

Newer cardiac markers, particularly cTnT and cTnI, are important diagnostic and prognostic markers for patients with ACS. These assays provide greatest benefit when cardiac marker results are coupled with treatment decisions and the application of proven and effective therapies to improve survival and prevent recurrent MI. Emerging data demonstrate that elevated concentrations of cardiac markers identify patients at greatest risk of death and infarction who benefit maximally from more potent therapies such as prolonged antithrombin therapy and platelet GP IIb/IIIa inhibition.

References

1. Hamm CW, Ravkilde J, Gerhardt W, et al. The prognostic value of serum troponin T in unstable angina. *N Engl J Med* 1992;327:146–150.
2. Lindahl B, Venge P, Wallentin L, for the FRISC Study Group. Relation between troponin T and the risk of subsequent cardiac events in unstable coronary artery disease. *Circulation* 1996;93:1651–1657.
3. Norgaard BL, Andersen K, Dellborg, et al. Admission risk assessment by cardiac troponin T in unstable coronary artery disease: Additional prognostic information from continuous ST segment monitoring. *J Am Coll Cardiol* 1999;33:1519–1527.
4. Pettijohn TL, Doyle T, Spiekerman AM, et al. Usefulness of positive tropo-

nin T and negative creatine kinase levels in identifying high-risk patients with unstable angina pectoris. *Am J Cardiol* 1997;80:510–511.

5. Hamm CW, Goldmann BU, Heeschen C, et al. Emergency room triage of patients with acute chest pain by means of rapid testing for cardiac troponin T or troponin I. *N Engl J Med* 1997;337:1648–1653.
6. Polanczyk CA, Lee TH, Cook F. Cardiac troponin I as a predictor of major cardiac events in emergency department patients with acute chest pain. *J Am Coll Cardiol* 1998;32:8–14.
7. Galvani M, Ottani F, Ferrini D, et al. Prognostic influence of elevated values of cardiac troponin I in patients with unstable angina. *Circulation* 1997;95: 2053–2059.
8. Antman EM, Tanasijevic MJ, Thompson B, et al. Cardiac-specific troponin I levels to predict the risk of mortality in patients with acute coronary syndromes. *N Engl J Med* 1996;335:1342–1349.
9. Ohman EM, Armstrong PW, Christenson RH, et al. Cardiac troponin T levels for risk stratification in acute myocardial infarction. *N Engl J Med* 1996;335:1333–1341.
10. Joint International Society and Federation of Cardiology, World Health Organization Task Force on Standardization of Clinical Nomenclature. Nomenclature and criteria for diagnosis of ischemic heart disease. *Circulation* 1979;59:607–608.
11. Kleiman NS, Granger CB, White HD, et al. Death and nonfatal reinfarction within the first 24 hours after presentation with an acute coronary syndrome: Experience form GUSTO-IIb. *Am Heart J* 1999;137:12–23.
12. Hamm CW, Heeschen C, Goldmann B, et al. Benefit of abciximab in patients with refractory unstable angina in relation to serum troponin T levels. *N Engl J Med* 1999;340:1623–1629.
13. Heeschen C, Hamm CW, Goldmann B, et al. Troponin concentrations for stratification of patients with acute coronary syndromes in relation to therapeutic efficacy of tirofiban. *Lancet* 1999;354:1757–1762.
14. Lindahl B, Venge P, Wallentin L, et al. Troponin T identifies patients with unstable coronary artery disease who benefit from long-term antithrombotic protection. *J Am Coll Cardiol* 1997;29:43–48.
15. Kong DF, Califf RM, Miller DP, et al. Clinical outcomes of therapeutic agents that block the platelet glycoprotein IIb/IIIa integrin in ischemic heart disease. *Circulation* 1998;98:2829–2835.
16. The CAPTURE Investigators. Randomised placebo-controlled trial of abciximab before and during coronary intervention in refractory unstable angina: The CAPTURE study. *Lancet* 1997;349:1429–1435.
17. Heeschen C, van den Brand MJ, Hamm CW, et al. Angiographic findings in patients with refractory unstable angina according to troponin T status. *Circulation* 1999;104:1509–1514.
18. The PRISM Study Investigators. A comparison of aspirin plus tirofiban with aspirin plus heparin for unstable angina. *N Engl J Med* 1998;338:1498–1505.
19. The PRISM-PLUS Investigators. Inhibition of the platelet glycoprotein IIb/IIIa receptor with tirofiban in unstable angina and non-Q-wave myocardial infarction. *N Engl J Med* 1998;338:1488–1497.
20. Hahn SS, Chae C, Giugliano R, et al. Troponin I levels in unstable angina/non-Q wave myocardial infarction patients treated with tirofiban, a glycoprotein IIb/IIIa antagonist. *J Am Coll Cardiol* 1998;31(suppl A):229A. Abstract.
21. Oler A, Whooley MA, Oler J, et al. Adding heparin to aspirin reduces

the incidence of myocardial infarction and death in patients with unstable angina: A meta-analysis. *JAMA* 1996;276:811–815.
22. FRISC Study Group. Low molecular weight heparin (Fragmin) during instability in coronary artery disease. *Lancet* 1996;347:561–568.
23. Ramanathan K, Stewart JT, Theroux P, et al. Admission troponin T may predict 90 minute TIMI flow after thrombolysis. *Circulation* 1997;96(suppl): I270. Abstract.
24. Hudson MP, White HD, Anderson RD, et al. Bedside risk stratification using both troponin T and ST-segment resolution in acute myocardial infarction: A GUSTO III substudy. *Circulation* 1998;98(suppl):I493. Abstract.

Chapter 10

The Evaluation of Acute Coronary Syndrome in the Emergency Department: *The Impact of Cardiac Biomarkers and ST-Segment Trend Monitoring*

David A. Grundy, MD and W. Brian Gibler, MD

Introduction

Despite great improvements in technology and triage protocols over the past decade, the diagnosis, treatment, and disposition of the patient presenting with chest pain or related symptoms suggestive of acute coronary syndrome (ACS) remain among the greatest challenges in emergency medicine. Approximately 8 million patients annually present to emergency departments (EDs) with chest pain or related complaints suggestive of ACS. Of these, 5 million are admitted for further work-up. Of the patients admitted, approximately 30% ultimately receive the diagnosis of ACS, and 300,000 (or 6%) die of short-term complications of infarction. A noncardiac diagnosis is assigned to more than 40% of the patients actually admitted.[1,2]

Perhaps of equal significance, of the 40% of patients with a suggestive complaint who are discharged to home, 1% to 2% will in fact have a missed acute myocardial infarction (AMI). These missed AMIs account for as much as 20% of malpractice settlements and awards involving emergency physicians.[3]

These figures suggest that although emergency physicians are capable of accurately diagnosing and treating ACS in the emergency set-

From: Adams JE III, Apple FS, Jaffe AS, Wu AHB (eds). *Markers in Cardiology: Current and Future Clinical Applications.* Armonk, NY: Futura Publishing Company, Inc.; © 2001.

ting, improvement can be realized through the application of a protocol-driven approach to the evaluation and treatment of these patients. As new treatment modalities such as glycoprotein (GP) IIb/IIIa inhibitors and low molecular weight heparins are developed, it is important to stratify patients with possible ACS on the basis of risk and to define optimum treatment regimens. This chapter addresses three areas of ongoing research with the potential to improve significantly the care of the patient with chest pain and possible ACS in the ED. First, this chapter summarizes recent advances in the use of serum markers of myocardial necrosis. An emphasis is placed on the role of serum markers in risk stratification of the patient with ACS. Second, the role of continuous electrocardiographic ST-segment trend monitoring in the detection of rest myocardial ischemia is reviewed. Finally, the use of ED-based chest pain centers as a cost-effective triage tool is discussed, with a focus on the University of Cincinnati and Mayo Clinic models that emphasize detection of myocardial necrosis and ST-segment trend monitoring.

Diagnosis of Myocardial Necrosis—Overview

An important focus in evaluating the spectrum of ACS patients is detecting the presence of myocardial necrosis as early as possible after the onset of symptoms. Differentiating non-Q-wave MI from unstable angina depends on the demonstration of necrosis. Evidence of ongoing necrosis may identify unstable patients who require a higher level of care, and may, in some instances, prompt early mechanical intervention such as with percutaneous transluminal coronary angioplasty (PTCA) or stent placement. At present, the detection of myocardial necrosis continues to rely on the detection of abnormally elevated serum levels of cardiac cytosolic proteins. The results of several large, prospective, clinical trials over the last 5 years have greatly clarified the diagnostic and prognostic value of the various serum markers in the patient with ACS. This chapter emphasizes the use of markers of myocardial necrosis in the emergency setting.

Myoglobin

Serum myoglobin has certain features that make it particularly useful as an early marker of myocardial necrosis, but with several important limitations. Myoglobin reaches peak concentration very rapidly after the onset of myocardial necrosis. Previous studies have shown detectable elevations of myoglobin as early as 1 hour after the onset of symptoms, and peak levels within 4 to 5 hours.[4] The measurement of serum myoglobin levels has been shown to have impressive sensitivity for myocardial necrosis.[5] Gibler and colleagues[4] have reported 1- and 3-hour sensitivities of 62% and 100%, respectively, marginally superior to the performance of the standard MB isoenzyme of creatine kinase (CK-MB).

The major limitation of myoglobin as a marker for myocardial necrosis is the relatively poor specificity of this protein. Current assays are not capable of distinguishing cardiac from skeletal muscle myoglobins. For this reason, the specificity of myoglobin for AMI at 3 hours after an index event is only approximately 80%, compared with over 90% for CK-MB.[6] Another disadvantage of this biomarker is that myoglobin is cleared by the kidneys, leading to potential false-positive results in patients with renal failure.[7] A final significant disadvantage to using myoglobin is that it has very rapid clearance kinetics. Winter and colleagues[8] demonstrate that although myoglobin has the highest negative predictive value of any of the cardiac markers at very early time-points, the negative predictive value falls sharply starting at approximately 6 hours. This is due to the rapid clearance of myoglobin from the serum after myocardial necrosis.

Despite important limitations, myoglobin is still useful as an early marker of AMI in the ED when used in conjunction with a more specific marker. The specificity of myoglobin can be improved by correlation with concurrent measurements of the skeletal muscle-specific enzyme carbonic anhydrase.[7]

CK-MB

CK-MB remains the most commonly used serum marker for the diagnosis of myocardial necrosis. Although not an ideal marker, CK-MB has several features that make it a useful tool in the evaluation of the patient with chest pain and possible ACS. The kinetics of CK-MB release allow for the sensitivity and specificity for myocardial necrosis to exceed 90% at 4 to 6 hours, and sensitivity that approaches 100% at 11 hours in the ED.[9] For this reason, the measurement of serial CK-MB levels remains the gold standard for the diagnosis of non-Q-wave MI. An abnormal level or an upward trend may prompt early intervention in the cardiac catheterization laboratory directly from the ED or triage of the patient to a higher level of care such as an intensive care unit. Elevated CK-MB levels also may have a role in the prediction of ischemic complication in ACS patients with an initially nondiagnostic electrocardiogram (ECG).[10]

Troponin

The cardiac troponins have enjoyed an increasingly important role in the evaluation of ACS in the ED. Cardiac troponin I (cTnI) and cardiac troponin T (cTnT) are both regulatory proteins associated with the contractile apparatus in cardiac and skeletal muscle. Together, they control the calcium-dependent interaction between actin and myosin and thereby the coordinated contraction of the muscle fibers. Sequence differences between cardiac and skeletal muscle troponins have allowed for the development of highly cardiac-specific assays for both proteins. A rapid point-of-care assay is available for cTnT that yields clinically reliable results in approximately 20 minutes.[11]

The kinetics of initial release of the troponins compare favorably with those of CK-MB. Both cTnI and cTnT begin to appear in the serum approximately 3 to 4 hours after the onset of myocardial necrosis. An initial peak of serum troponins is typically observed by 14 hours after the onset of symptoms. This is often followed by a second, smaller peak 3 to 5 days after an MI.[12]

In contrast to CK-MB, the cardiac troponins exist in two distinct compartments in myocytes. The first is a small cytosolic pool that is released early during necrosis. A significantly larger fraction, however, is associated with the actin/myosin contractile apparatus itself. It is the slow liberation of this component as the contractile apparatus breaks down that keeps troponin levels elevated for 7 to 11 days after AMI.[13]

In a recent ED-based study, the sensitivity of the cardiac troponins for AMI was similar to that of CK-MB at all time-points up to 24 hours. At 24 hours, the sensitivity of cTnT for AMI was 97% and the specificity was 92%.[14]

Data derived from several large, multicenter, prospective studies have demonstrated the prognostic value of the cardiac troponins in patients with possible ACS. Data from the GUSTO-IIa, TIMI-IIIb, and FRISC studies show that elevated troponin levels within the first 24 hours after the onset of chest pain have a clear association with adverse cardiac events.[15–17] Until quite recently, however, it has not been clear what role troponin levels should play in risk stratification of the large and heterogeneous group of patients presenting to EDs without clear evidence of ACS and an initially nondiagnostic ECG.

Several recent studies have helped to define this role. Sayre and colleagues,[14] in a prospective study of 667 patients presenting to the ED with chest pain and nonspecific ECGs, have shown that cTnT levels greater than 0.2 μg/L predict a 3.5-fold elevated risk of cardiac complications within 60 days. Furthermore, the prognostic value of an elevated troponin level in this study was independent of the CK-MB results. These results supplement and confirm those of a prospective trial by Hamm and colleagues[18] in which a similar patient population of 773 patients from the ED presenting with chest pain had serial cTnT and cTnI levels recorded and correlated with 30-day complication rates and final diagnoses. This study demonstrated a negative predictive value of 98.9% for cTnT and 99.7% for cTnI for AMI. Interestingly, 36% of patients with a final diagnosis of unstable angina also had elevated troponin levels, whereas only 5% of these patients had elevated CK-MB levels. These individuals likely represent a high-risk subset of patients with microinfarctions or minimal myocardial injury who would have been missed by CK-MB screening alone.[18]

As the development of new treatment methods tailored to the various subsets of ACS patients continues, such as the use of GP IIb/IIIa inhibitors in patients with unstable angina and non-Q-wave MI, the use of cardiac markers for early detection of necrosis and risk stratification of ED patients is becoming increasingly important. Underscoring this point, in the recent CAPTURE trial, only the group of patients with unstable angina and an elevated cTnT level derived benefit from treatment with abciximab, a GP IIb/IIIa inhibitor.[19]

Recent recommendations of the National Academy of Clinical Biochemistry provide a guide to the appropriate use of cardiac markers for the triage of patients with possible ACS. These recommendations suggest the use of two markers for all patients evaluated for ACS. The first or "early" marker must reliably become elevated very early (within 6 hours) after ischemic myocardial damage. The relative release kinetics of the currently available serum markers would seem to favor myoglobin in this role.

The second marker is the "definitive marker," upon which a diagnosis of AMI can be definitively made or excluded. This marker must have a high sensitivity and specificity for AMI and become elevated between 6 and 9 hours after the onset of ischemia. The marker must stay elevated for several days after the event. Because of excellent sensitivity and specificity, prolonged release kinetics, and utility in risk stratification for ACS, the troponins are clearly the best suited cardiac biomarkers for this role. It is anticipated that cTnI or cTnT will ultimately replace CK-MB as the standard marker for the diagnosis of non-Q-wave MI.[20] Serial measurement of both CK-MB and troponin in the evaluation of the suspected ACS patient remains the preferred strategy. Elevated troponin levels also define a subset of patients with nondiagnostic ECGs that was previously missed by CK-MB-based "rule-out MI" strategies. These patients should benefit from admission to intensive care settings and more aggressive treatment such as GP IIb/IIIa platelet inhibitors and low molecular weight heparins.

Diagnosis of Rest Ischemia

Although current management strategies for unstable angina and non-Q-wave MI are evolving, it is understood that patients presenting with evidence of ST-segment elevation on the initial 12-lead ECG represent a high-risk subset that derives maximum benefit from emergency reperfusion.[21] It has also been clearly demonstrated that time to reperfusion has a direct impact on outcome. The initial ECG, however, only shows a clear injury pattern in approximately 50% of all patients presenting with AMI.[22] Up to 20% of these patients without ST-segment elevation on presentation will eventually develop ECG changes of Q waves consistent with transmural injury over the next 24 to 48 hours.[23] In this period before the development of definitive ECG criteria, time may be lost during which reperfusion therapy could provide maximum benefit.

One of the reasons that a single 12-lead ECG is inadequate to fully evaluate the ACS patient is that the progression of coronary occlusion is an intrinsically dynamic process. Thrombosis is the outcome of a complex interaction of endogenous thrombogenic and thrombolytic pathways. Preceding complete vessel occlusion, there may be a relatively prolonged period in which near normal vessel patency alternates with near complete occlusion after rupture of an atheroscleromatous plaque. It is a chance occurrence in some cases of unstable angina for a randomly acquired ECG to capture an episode of ischemia or the injury pattern of ST-segment elevation. Even when serial, standard, 12-lead ECGs are acquired whenever the patient complains of pain, a partial occlusion may temporarily resolve in the time required to obtain an ECG. This strategy of obtaining a 12-lead ECG when a patient complains of pain also fails to detect ischemia in patients with silent or nonpainful ischemia or in those who are unable or unwilling to complain of a change in their symptoms. Because morbidity and mortality are similar in patients with silent ischemia and in those with classic symptoms, early detection is critically important in both groups of patients with ACS.[24] Serial ECGs can be acquired on some fixed schedule, such as every 30 minutes, but there is no consensus on an optimum frequency of acquisition that gives the best outcome without placing an overwhelming burden on nursing and ancillary staff.

Continuous ST-segment trend monitoring, however, provides a means of identifying the onset of ischemia at rest at the earliest possible moment for the subset of patients having ACS. Trend monitoring, performed by computer algorithm, is commonly used to detect early reocclusion after catheter-based interventions of the coronary arteries. Trend monitoring has also been used in the detection of intraoperative and postoperative myocardial ischemia.[25] A logical extension of the use of ST-segment trend monitoring is in patients being evaluated for possible ACS in the ED or in ED-based chest pain units (CPUs). In this setting, the earliest possible detection of rest ischemia can potentially have the greatest impact on patient outcome by initiating emergency reperfusion therapies. This is reflected in the design of the Heart ER program at the University of Cincinnati and those programs modeled after it. In this program, low- to moderate-risk patients being evaluated for possible ACS have continuous ST-segment monitoring performed and serial 12-lead ECGs obtained for a 6-hour period before exercise testing is performed at the end of the observation period.[26] This strategy provides the highest possible sensitivity for rest ischemia in both symptomatic and asymptomatic patients.

The majority of studies of ST-segment trend monitoring have been performed in the setting of intensive care units. In a study of patients with AMI during angiography, Krucoff and colleagues[27] demonstrated that ST-segment-based determination of vessel patency was 90% sensitive and 92% specific compared with an angiographic determination of patency. In this trial, achievement of an ST-segment steady state was found to have a 100% sensitivity and specificity for subtotal rather than total coronary occlusion.

ST-segment trend monitoring has also been shown to be a useful method for risk stratification in patients with unstable angina, providing prognostic information in addition to the initial ECG and patient clinical characteristics.[28] In a prospective analysis of 212 patients admitted with unstable angina, Patel and colleagues[29] found that the presence of transient ischemia on ST-segment trend monitoring was an independent, highly significant predictor of subsequent death or nonfatal MI. Patients with total duration of ischemia less than 60 minutes per 24 hours had a 2.7 times increased risk of death or MI, whereas those with greater than 60 minutes per 24 hours of transient ischemia had an 8 times greater risk. The majority of these adverse events occurred within the first 6 weeks after admission, with one third occurring in the hospital.[29] Several recent case reports address the clinical impact of ST-segment monitoring, in which this technology detected asymptomatic ischemia and motivated early treatment with thrombolytics or identified ST-segment instability preceding cardiac arrest.[30,31]

Fesmire and colleagues[22] have recently extended this work to the evaluation of patients in the ED. In a prospective, observational study in which 1000 patients were enrolled, the sensitivity of ST-segment trend monitoring was found to be clearly superior to that of a single initial ECG for detection of all AMIs (68.1% versus 55.4%) and of other ACS (34.2% versus 27.5%).[22] The authors note that trend monitoring gives a relative increase of 35.5% in detection of myocardial injury in this population. Having ST-segment trend data routinely available in

EDs and chest pain centers could increase substantially the number of candidates for emergency reperfusion. Although additional large, randomized, prospective ED-based studies are needed to define clearly the costs and benefits of this technology, continuous ST-segment monitoring has a potentially important role in the care of the patient with ACS in the setting of ED-based chest pain centers.

Chest Pain Centers

The past two decades have seen an explosive growth in the number of ED-based chest pain centers. First implemented at St. Agnes Hospital in Baltimore in 1981, currently there are an estimated 1000 such units in the US. Several different protocols are in use today, but two model programs have excellent records for safety and cost effectiveness. The Heart ER program at the University of Cincinnati and the Mayo Clinic Chest Pain Center, which was modeled in part on the Cincinnati program, use serial measurements of serum biomarkers for detection of myocardial necrosis and ST-segment trend monitoring for detection of rest ischemia as integral aspects of a protocol-driven approach to patients with suspected ACS.

The clinical pathway used at the University of Cincinnati Hospital for the evaluation of patients with chest pain is shown in Figure 1.

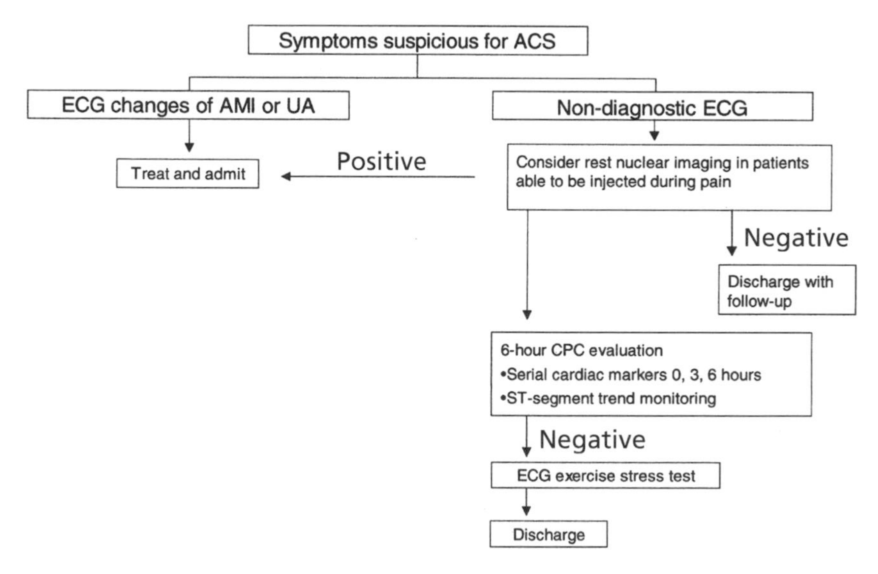

Figure 1. University of Cincinnati "Heart ER" strategy. ACS = acute coronary syndrome; AMI = acute myocardial infarction; CPC = chest pain center; ECG = electrocardiogram; UA = unstable angina.

More than 2500 patients of low to intermediate risk for ACS have been evaluated in this unit from October 17, 1991 to the present. Patients over 25 years of age are eligible for Heart ER evaluation. Exclusion criteria are ST-segment elevation on initial ECG, ongoing chest pain in patients with a clinical presentation consistent with unstable angina, and signs of hemodynamic instability.

Evaluation methods are protocol-driven. Serial CK-MB, myoglobin, and cTnT levels are determined at 0, 3, and 6 hours. Continuous ST-segment trend monitoring is performed throughout the evaluation period. If all these tests are normal after a 6-hour period of observation, the patient is evaluated for exercise-induced ischemia with graded exercise testing (GXT) or radionuclide scanning. If the GXT is normal, the patient is discharged from the ED with provision for close follow-up by the patient's private physician or a chest pain clinic. In this protocol, there is also the provision for early Tc-99 sestamibi single photon emission computed tomography scanning for patients with active chest pain.

To date, the Heart ER program has an excellent record for the efficient and safe evaluation of patients with potential ACS. Of 2131 patients reviewed in 1998, 1822 (85.5%) were discharged from the ED, and 309 (14.5%) were admitted. Of the patients admitted, 94 (30%) were found to have a cardiac cause for their chest pain. There were 23 patients with AMI and 63 cases of unstable angina or angina identified. Of the patients discharged, follow-up was obtained in 1696 (93%). The rate of cardiac events in this group was only 0.53% (7 PTCA, 1 coronary artery bypass graft, 1 death).[32]

The "Chest Pain Unit" at the Mayo Clinic in Rochester, Minnesota is very similar to the University of Cincinnati model. This unit is a 4-bed area of the ED with dedicated full-time nursing. The clinical pathway used is shown in Figure 2. Patients considered to be at low to intermediate risk for ACS are evaluated in this unit. Between November 1995 and March 1997, a community-based, prospective, randomized trial was performed that compared the outcomes and cost of care of patients evaluated in the CPU to patients admitted to the cardiology service for routine diagnosis and care. The study found no significant difference in the rate of adverse cardiac events between the two groups. Cost of care, however, was significantly less in the CPU group.[33]

In conclusion, both of these highly successful clinical pathways have mechanisms for the detection of myocardial necrosis through serial determination of a panel of serum cardiac biomarkers, rest ischemia by continuous ST-segment trend monitoring and serial ECGs, and exercise-induced ischemia by provocative testing after a 6- to 9-hour observation period. The data thus far confirm that this approach to the patient with chest discomfort and possible ACS provides excellent

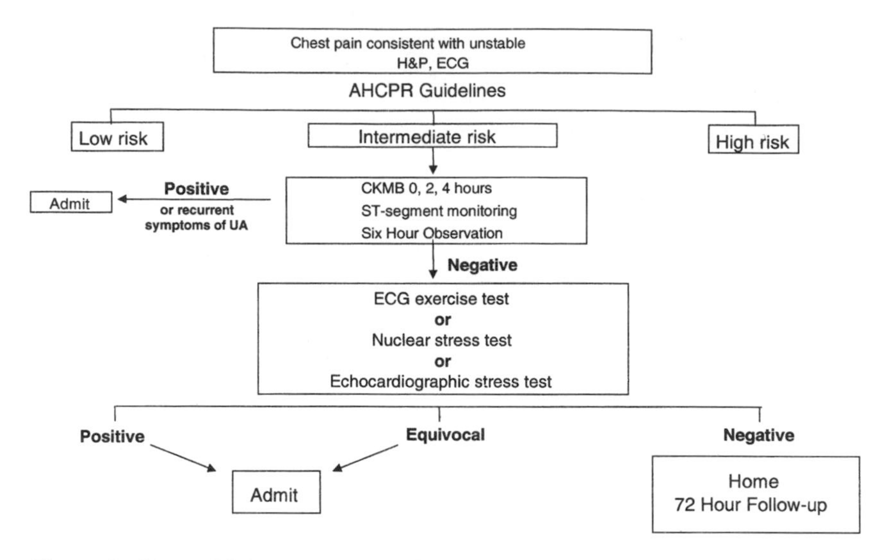

Figure 2. Mayo Clinic strategy. AHCPR = Agency for Health Care Policy and Research; CKMB = creatine kinase MB; ECG = electrocardiogram; H&P = history and physical; UA = unstable angina.

diagnostic sensitivity and patient safety while significantly improving the cost effectiveness of care.

References

1. American Heart Association. *Heart and Stroke 1998 Statistical Update, American Heart Association.* 1998.
2. Jesse RL, Kontos MC. Evaluation of chest pain in the emergency department. *Curr Probl Cardiol* 1997;22:149–236.
3. McCarthy BD, Beshansky JR, Agostino RB et al. Missed diagnosis of acute myocardial infarction in the emergency department: Results from a multicenter study. *Ann Emerg Med* 1994;22:579–582.
4. Gibler WB, Gibler CD, Weinshenker E, et al. Myoglobin as an early indicator of acute myocardial infarction. *Ann Emerg Med* 1987;16:851–856.
5. Polanczyk CA, Lee TH, Cook EF, et al. Value of additional two-hour myoglobin for the diagnosis of myocardial infarction in the emergency department. *Am J Cardiol* 1999;83:525–529.
6. Storrow AB, Gibler WB. The role of cardiac markers in the emergency department. *Clin Chim Acta* 1999;284:187–196.
7. Vuori J, Huttunen K, Voutikka P, Vaananen HK. The use of myoglobin/carbonic anhydrase III ratio as a marker for myocardial damage in patients with renal failure. *Clin Chim Acta* 1997;265:33–40.
8. Winter RJ, Koster RW, Sanders GT. Value of myoglobin, troponin T, and CK-MB mass in ruling out an acute myocardial infarction in the emergency room. *Circulation* 1995;12:3401–3407.

9. Gibler WB, Lewis LM, Erb RE, et al. Early detection of acute myocardial infarction in patients presenting with chest pain and non-diagnostic ECGs: Serial CK-MB sampling in the emergency department. *Ann Emerg Med* 1990; 19:1359–1366.
10. Hoekstra JW, Hedges JR, Gibler WB, et al. Emergency department CK-MB – a predictor of ischemic complications. *Acad Emerg Med* 1994;1:17–28.
11. REACTT Investigators Study Group. Evaluation of a bedside whole-blood rapid troponin T assay in the emergency department. *Acad Emerg Med* 1997; 4:1018–1024.
12. Mair J. Progress in myocardial damage detection: New biochemical markers for clinicians. Crit Rev Clin Lab Sci 1997;34:1–66.
13. Ravkilde J, Nissen H, Horder M, et al. Independent prognostic value of serum creatine kinase isoenzyme MB mass, cardiac troponin T, and myosin light chain levels in suspected acute myocardial infarction. *J Am Coll Cardiol* 1995;25:574–581.
14. Sayre MR, Kaufmann KH, Gibler WB, et al. Measurement of cardiac troponin T is an effective method for predicting complications among emergency department patients with chest pain. *Ann Emerg Med* 1998;3:539–549.
15. Ohman EM, Armstrong PW, Topol EJ, et al. Cardiac troponin T levels for risk stratification in acute myocardial ischemia. *N Engl J Med* 1996;335: 1333–1341.
16. Antman EM, Tanasijevic MJ, Braunwald E. Cardiac-specific troponin-I levels to predict the risk of mortality in patients with acute coronary syndromes. *N Engl J Med* 1996;335:1342–1349.
17. Lindahl B, Venge P, Wallentin L, FRISC study group. Relation between troponin T and the risk of subsequent cardiac events in unstable coronary artery disease. Circulation 1996;93:1651–1657.
18. Hamm CW, Goldmann BU, Meinertz T. Emergency room triage of patients with acute chest pain by means of rapid testing for cardiac troponin T or troponin I. *N Engl J Med* 1997;337:1648–1653.
19. Hamm CW, Heeschen C, Goldmann B, et al. Benefit of abciximab in patients with refractory unstable angina in relation to serum troponin T levels. *N Engl J Med* 1999;340:1623–1629.
20. Wu AHB, Apple FS, Gibler WB, et al. National Academy of Clinical Biochemistry Standards of Laboratory Practice: The use of cardiac markers in coronary artery diseases. *Clin Chem* 1999;45:1104–1121.
21. Ryan TJ, Anderson JL, Antman EM, et al. ACC/AHA guidelines for the management of patients with acute myocardial infarction: A report of the American College of Cardiology/American Heart Association Task Force on Practice Guidelines (Committee on Management of Acute Myocardial Infarction). *J Am Coll Cardiol* 1996;28:1328–1428.
22. Fesmire FM, Percy RF, Calhoun FB, et al. Usefulness of automated serial 12-lead ECG monitoring during the initial emergency department evaluation of patients with chest pain. Ann Emerg Med 1998;31:3–11.
23. Gibler WB, Young GP, Hedges JR, et al. Acute myocardial infarction in chest pain patients with non-diagnostic ECGs: Serial CK-MB sampling in the emergency department. Ann Emerg Med 1992;21:504–512.
24. Myerburg RJ, Kessler KM, Mallon SM. Life-threatening ventricular arrhythmias in patients with silent myocardial ischemia due to coronary artery vasospasm. N Engl J Med 1992;326:1451–1455.
25. Leung JM, Voskanian A, Bellows WH. Automated electrocardiographic ST-segment trending monitors: Accuracy in detecting myocardial ischemia. Anesth Analg 1998;87:4–10.

26. Gibler WB, Runyon JP, Levy RC, et al. A rapid diagnostic and treatment center for patients with chest pain in the emergency department. *Ann Emerg Med* 1995;25:1–8.
27. Krucoff MW, Croll MA, Pope JE, et al. Continuously updated 12-lead ST-segment recovery analysis for myocardial infarct artery patency assessment and its correlation with multiple simultaneous early angiographic observations. *Am J Cardiol* 1993;71:145–151.
28. Patel DJ, Holdright DR, Fox KM, et al. Early continuous ST segment monitoring in unstable angina: Prognostic value additional to the clinical characteristics and the admission electrocardiogram. *Heart* 1996;75:222–228.
29. Patel DJ, Knight CJ, Fox KM, et al. Long-term prognosis in unstable angina. The importance of early risk stratification using continuous ST segment monitoring. *Eur Heart J* 1998;19:240–249.
30. Fu GY, Joseph AJ, Antalis G. Application of continuous ST-segment monitoring in the detection of silent myocardial ischemia. *Ann Emerg Med* 1994; 23:1113–1115.
31. Fesmire FM, Bardoner JB. ST-segment instability preceding simultaneous cardiac arrest and AMI in a patient undergoing continuous 12-lead ECG monitoring. *Am J Emerg Med* 1994;12:69–76.
32. Storrow AB, Gibler WB, Brennan T, et al. An emergency department chest pain rapid diagnosis and treatment unit: Results from a six-year experience. *Circulation* 1998;98(suppl):I425. Abstract.
33. Farkouh ME, Smars PA, Sherine EG, et al. A clinical trial of a chest-pain observation unit for patients with unstable angina. *N Engl J Med* 1998;339: 1883–1888.

Chapter 11

Cardiac Troponin T in Coronary Artery Disease: *Where Do We Stand?*

Evangelos Giannitsis, MD, Britta Weidtmann, MD, Margit Müller-Bardorff, MD, Norbert Frey, MD, and Hugo A. Katus, MD

Introduction

When we started our development of diagnostic assays for cardiac myofibrillar proteins in 1978, we hypothesized that the use of a highly abundant cardiospecific molecule as a protein marker for myocyte integrity may translate to the enhanced performance of biochemical tools for the diagnosis of myocardial injury.[1–4] With the development of an enzyme-linked immunosorbent assay (ELISA) for cardiac troponin T (cTnT),[5–7] these efforts resulted in a clinically useful assay suitable for testing the above hypothesis in single-center studies[8,9] and then in independent prospective multicenter trials.[10–14] Over time, the initial assay formats have been improved to better fit clinical needs.[15–18] This chapter attempts to characterize the available analytical systems, to summarize the clinical data on the diagnostic and prognostic value of cTnT elevations, and to indicate still unresolved issues of cTnT testing.

Analytical Systems

The cTnT assay exists in two formats: a test-strip, whole-blood assay for bedside testing (point of care)[15] and an electrochemiluminescence plasma or serum assay (ELECSYS, Roche Diagnostics, Indianapolis, IN)[19] for central laboratory measurement. Both assay formats use

From: Adams JE III, Apple FS, Jaffe AS, Wu AHB (eds). *Markers in Cardiology: Current and Future Clinical Applications.* Armonk, NY: Futura Publishing Company, Inc.; © 2001.

the same two monoclonal mouse antibodies that bind to the rod portion of cTnT on two antigenic determinants separated by 11 amino acid residues. Although initially one antibody was cardiospecific and the second was cross-reactive with skeletal troponin T,[7] the antibodies used in the current assays reveal less than 0.1% cross-reactivity with skeletal troponin T.[16] One assay improvement was the replacement of bovine cTnT, the protein standard in the early assays, with recombinant human cTnT.[19] Since the different assay formats contain the same antibodies and are calibrated to the same standard material, the different troponin T assays reveal a high interassay reproducibility.

The bedside test-strip assay may be analyzed by the physician or nurse on duty and is classified as positive if two lines (a control and a detection line) appear in the reading zone within 20 minutes after application of 150 μL of heparinized whole blood. The detection limit of this qualitative, dichotomized ("positive or negative") device is 0.1 μg/L. Only the time delay from blood application to the appearance of a positive line in the reading zone and the intensity of color of the test line indirectly indicate the levels of cTnT in blood.[15,20–22] The training status of the ward staff, ambient light conditions, and a number of other variables may affect the reproducibility of visual assessment of the rapid assay result, particularly if cTnT levels in blood are close to the detection limit of the device. Despite these shortcomings, in-field studies confirmed the clinical utility of this device for diagnosis of acute myocardial infarction (AMI) and for risk stratification of chest pain patients. In the study of chest pain patients by Hamm and coworkers,[23] the field testing of the cTnT whole-blood, rapid assay was equally efficient as the centralized batch analysis of cardiac troponin I (cTnI) measured from precentrifuged plasma samples. However, when Lindahl and coworkers[24] compared the results of cTnT measurements from qualitative bedside testing of whole blood with results from quantitative central laboratory measurement of plasma samples by the ELECSYS method, they found a superior performance of the quantitative assay. They recommended central laboratory testing as the preferred analytical measurement for the monitoring of patients with unstable angina.[24] To eliminate the variability induced by visual assessment of the presence or absence of a faint color line on the rapid assay, a reading device was developed for quantitative assessment of the test-strip results. This Cardiac Reader™ (Roche Diagnostics) is based on a charge coupled device (CCD) camera that measures the intensity of the positive line by light remission that can be calibrated and converted into quantitative concentrations by a standardized algorithm. The time required for the completion of a test-strip analysis by the reader is less than 20 minutes, and a signal lights up as soon as a positive line becomes detectable. The detection limit achieved with this device is 0.1

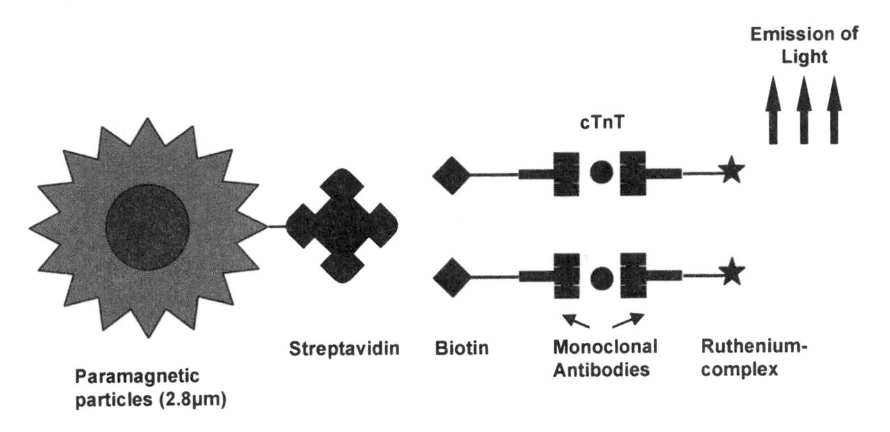

Figure 1. Components and test principle of the ELECSYS-Troponin Assay using ruthenium-labeled antibodies. Measurable light emission is secondary to activation of ruthenium and proportional to antigen concentrations.

µg/L, with a measuring range to 3.0 µg/L and a coefficient of variation of 9.0% to 13.6%, depending on cTnT concentrations.[18] In patients with AMI, similar levels are obtained by the reader and ELISA method at different points after the onset of symptoms, thus indicating comparable recoveries of cTnT fragments from blood by both methods.

In the early stages of test development, cTnT was measured by ELISA, which required a turnaround time of 45 minutes.[16] Furthermore, the Enzymun-Test system ES22 analyzer (Roche Diagnostics) was not suited for random access of test samples. To make cTnT determinations more feasible in emergency testing and clinical practice, a novel analyzer (ELECSYS) was developed, and the components of the cTnT assay were further improved. The ELECSYS analyzer uses an electrochemiluminescence method for detection (Fig. 1). This analyzer determines the changes of electronically activated luminescence of the second monoclonal antibody labeled with ruthenium. Activation is induced by displacement of the first antibody, which is conjugated to magnetic beads. This approach shortens the turnaround time of the assay to 12 minutes, while gaining markedly in sensitivity. Since the ELECSYS analyzer provides random access of samples, this instrument is suitable for emergency testing of chest pain patients. In addition to the adaptation of the cTnT assay to the ELECSYS analyzer, bovine cTnT was also replaced by recombinant human cTnT. The use of human[19] instead of bovine[17] cTnT led to the better recognition of the protein standard by the anti-human mouse antibodies, resulting in a shift of the standard curve and a decrease of cTnT concentrations, particularly in samples with high cTnT levels. Thus, due to the use of bovine standards, previous assays had led to an overestimation of circulating cTnT levels in patients with massive MI. When a method comparison was performed within a range of up to 25 μg/L in 999 samples, regression analysis disclosed $y = 0.06 + 0.62x$, $r = 0.988$. Disproportionate results between assays using human and bovine standard material were not observed in the lower range (0 to 0.2 μg/L); a good agreement between both methods was found $y = 0.001 + 1.00x$, $r = 0.984$, $n = 286$.

The improved analytical technology of the analyzer and the implementation of human standard material have led to the improved diagnostic performance of the ELECSYS assay at the detection limit. These technical improvements translate into the superior diagnostic performance of the assay, particularly at the lower concentration range, with a coefficient of variation of 3.9% at 0.1 μg/L. The measuring range of the assay extends from 0.01 to 30 μg/L, 0.03 μg/L corresponding to +3 standard deviations above mean normal blood levels (99% limit). The 0.03 μg/L discriminator value is a marked reduction of the upper limit of normal as compared with the 0.1 μg/L achieved with former assay formats. When 1268 blood samples from healthy individuals were tested with this assay, only five gave detectable measurements in the range of less than 0.021. Similar results were obtained when blood samples of patients without cardiac diseases were analyzed. Thus, it must be concluded that even with the more sensitive assays, normal levels of cTnT are not detectable, and elevations of cTnT below the clinical discriminator 0.1 μg/L must reflect protein marker release from injured myocardium.

Diagnostic Challenges

The diagnosis of AMI relies on the detection of cardiac markers in blood or their elevation above predefined normal limits. For years, myocardial muscle creatine kinase isoenzyme (CK-MB) elevations were regarded as the gold standard for the biochemical diagnosis of MI and myocardial cell injury. With the advent of more sophisticated analytical methods and more suitable marker molecules, discrepant results of these novel assays to CK-MB were observed. Thus, patients were reported with elevations of cTnT even in the absence of significant CK-MB levels in blood.[13] These findings have questioned the value of either CK-MB or the novel troponins for the diagnostic classification of chest pain patients, depending on the acceptance of the respective marker as the gold standard.

Comparative analysis of the release of cardiac markers in patients undergoing radiofrequency ablation and in patients with unstable angina revealed a greater increase in blood levels of cTnT than of CK-

MB and a more prolonged elevation after the index event, even when the conservative 0.1 μg/L cTnT discriminator value was used. Thus, in experimental models of cardiac injury, the relative increases of cTnT in blood were markedly higher than those of cardiac enzymes.

The diagnostic problem of using an even more sensitive cTnT assay format with a detection limit of 0.04 μg/L can be exemplified by a comparative analysis of CK-MB and cTnT in chest pain patients. Gerhardt and coworkers[25] found a continuous spectrum of corresponding values of CK-MB and cTnT. These values could be dichotomized according to the current discriminator values of CK-MB (>10 μg/L) and cTnT (>1.0 μg/L). These investigators found that out of 1577 ischemic patient episodes, 754 episodes were classified as AMI as suggested by concordant CK-MB greater than 10 μg/L and cTnT greater than 0.1 μg/L, 93 episodes were classified as minimal myocardial damage as suggested by cTnT greater than 0.1 μg/L but normal CK-MB, and 730 episodes had neither AMI nor minor myocardial damage as suggested by normal CK-MB and cTnT. Below the discriminator value, patients with coronary artery disease tended to have higher cTnT levels (mean 0.07 μg/L) than those of patients without obvious coronary artery disease (mean 0.015 μg/L). This finding indicates that even in patients classified as having stable coronary disease, cardiac markers may be released from myocardium. The clinical and prognostic implications of subtle cTnT elevations below 0.1 μg/L must be investigated in prospective trials.

Since the release of cardiac markers is not indicative of the inflicting cause, a diagnosis of AMI can be pursued only if there are clinical findings of ischemia such as anginal pain or electrocardiographic changes. However, there are also nonischemic damages to the heart such as myocarditis, toxic damage, cardiac transplant rejection, and heart contusion, when elevations of cTnT in blood have been reported.[25–29]

We have investigated in detail one of the groups of patients in whom cTnT elevations were present in the absence of severe coronary artery disease. In a prospective trial, cTnT elevations were observed in 18 of 58 patients with acute pulmonary embolism.[30] The diagnosis of acute pulmonary embolism was based on a standard diagnostic workup, including the assessment of clinical probability. The cTnT-positive patients revealed right ventricular dysfunction, hypotension, shock, and death more often than did the cTnT-negative patients (Fig. 2). The cTnT status was the best predictor of a patient's risk, based on multivariate analysis, and cTnT performed better than the angiographic classification of severity of pulmonary embolism. The release of cTnT was not due to coexisting coronary artery disease since cTnT was equally prevalent in patients with and without significant coronary artery

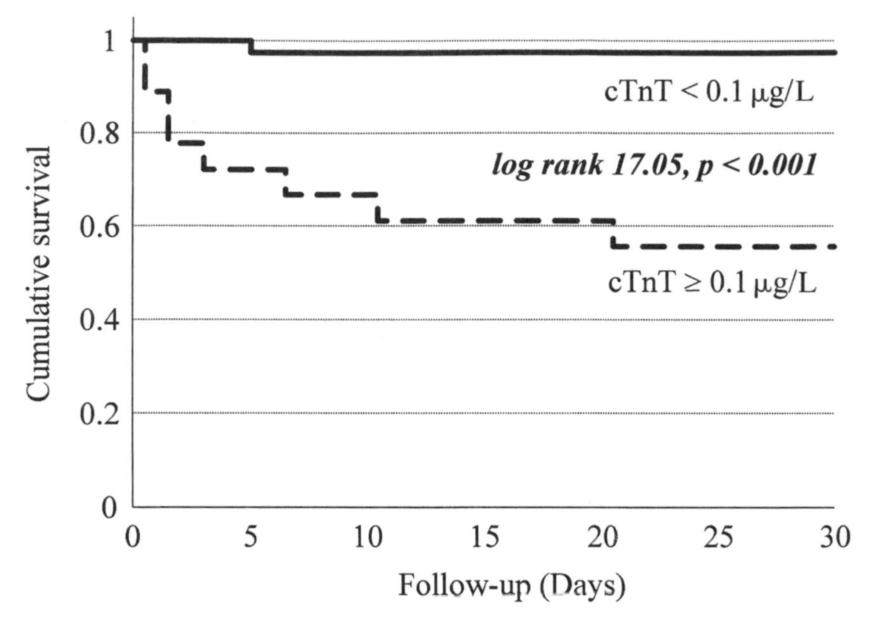

Figure 2. Cumulative survival plot (Kaplan-Meier method) in 58 patients with confirmed pulmonary embolism stratified by cardiac troponin T (cTnT) values.

narrowing. Thus, it is likely that severe hypoxemia in combination with right ventricular distension may lead to the leakage of intracellular molecules.

Prognostic Implications

In patients with ST-segment elevation AMI, cardiac markers may serve to confirm the clinical diagnosis, to monitor infarct evolution, to test the efficiency of reperfusion therapy, and to estimate infarct size. In recent years, an additional indicator of AMI risk has been identified—elevated levels of cTnT.

Stubbs and coworkers[31] reported on a threefold higher mortality rate in patients arriving at the hospital with an already elevated cTnT versus patients who were still negative for cTnT. These findings confirmed results of the GUSTO-IIa study, where in the ST-elevation group, the 30-day mortality rate was 4.7% in cTnT-negative patients and 13% in cTnT-positive patients.[13] From these two studies, factors contributing to the higher mortality of the cTnT-positive patients with ST-segment elevation AMI were not readily evident. However, the data by Stubbs and coworkers were confirmed by the GUSTO-III investigators.[32] These investigators prospectively analyzed 12,806 patients undergoing thrombolysis for ST-segment elevation AMI. They also reported on a significantly higher mortality rate of cTnT-positive patients upon admission versus patients who were negative for cTnT during the hospital stay, and at up to 6 months and 12 months of follow-up. In these trials, among cTnT-positive patients, mortality rates were not related to the elapsed time from onset of symptoms to the initiation of therapy. There is some recent evidence that patients who are positive for cTnT on admission may be more resistant to thrombolytic treatment and achieve complete, normal reperfusion (TIMI 3 flow) of the infarct-related artery less frequently than patients who are negative for cTnT.[33,34]

To test whether a positive cTnT on admission remains predictive of patients' outcome, even when the effectiveness of target vessel recanalization is controlled by angiography, we prospectively investigated patients with ST-segment elevation AMI treated by primary percutaneous transluminal coronary angioplasty and provisional stenting.[35] This trial included only patients with successful direct percutaneous transluminal coronary angioplasty, and thus does not explain the different 30-day mortality rates of 16.1% versus 4.4% observed in cTnT-positive compared with cTnT-negative patients (Table 1). However, we found that the cTnT-positive patients arrived at the hospital after the onset of symptoms significantly later than did the cTnT-negative patients (3.85 hours versus 2.35 hours). Thus, it is likely that recanalization of

Table 1

Findings in 183 Patients with Acute ST-Segment Elevation AMI after Successful Direct PTCA in Relation to Admission cTnT

	cTnT <0.1 (n = 90)	cTnT >0.1 (n = 93)	*P* Value
Mean delay from onset of symptoms (h) (range)	2.35 (0.5–10)	3.85 (0.5–8)	<0.01
LV ejection fraction <0.35	8.9%	26.9%	<0.01
Cardiac death at 30 days	4.4%	16.1%	0.01
Cardiac death at 12 months	5.6%	17.2%	0.01
Combined endpoint at 12 months*	17.8%	36.6%	<0.01

Data are given as relative frequencies or as mean values. *Death, nonfatal MI, target vessel reintervention. MI = myocardial infarction; AMI = acute myocardial infarction; PTCA = percutaneous transluminal coronary angioplasty; cTnT = cardiac troponin T; LV = left ventricular.

the infarct vessel is less effective in terms of myocardial salvage or microvascular reperfusion in the cTnT-positive patients. Furthermore, cTnT-positive patients had poorer left ventricular function and more severe coronary artery disease than did the cTnT-negative patients, although estimates of infarct size were not significantly different between both groups. Our findings are in line with a recent report of the FRISC study group.[33] They reported a mortality rate of 25% versus 4.5% for cTnT-positive versus cTnT-negative patients on admission with ST-segment elevation AMI, despite infarct sizes that were comparable as estimated by cTnT levels at 72 hours. The FRISC authors also noted that cTnT-positive patients arrived later to the hospital after the onset of symptoms (4.2 versus 2.0 hours; $P<0.001$), had less frequently complete reperfusion of the infarct-related artery (TIMI 3 flow) after thrombolysis, and suffered more frequently repetitive angina episodes during their AMI course than did cTnT-negative patients. Thus, it is obvious from many trials that a positive cTnT value on admission identifies a high-risk subgroup of patients with ST-segment elevation AMI, which apparently is not efficiently treated by recanalization therapy.

In many prospective trials on patients with acute coronary syndromes but without ST-segment elevation, it has also been shown that a positive cTnT on admission or within the first 24 hours after admission identifies a high-risk subset of patients.[10–13] These cTnT-positive, unstable angina patients have an acute and long-term cardiac event rate that is not different from that of patients with reperfused AMI, although a normal CK-MB value excluded the presence of AMI in this group. Notably, the increase in cardiac risk is not limited to the index hospital stay but persists for at least 28 months after the index event.[36] The factors contributing to the increased long-term mortality rate after an unstable angina episode resulting in positive cTnT are not yet understood.

Recent investigations have indicated that plaque instability may determine long-term risk in unstable angina. Plaque instability is, among other variables, dependent on local and systemic inflammation, leading to activation of tissue metalloproteinases and increased blood cell adhesion and transmigration.[37–39] In clinical trials on unstable angina patients, it was observed that levels of biochemical indices of inflammation in blood such as C-reactive protein, serum amyloid A, and fibrinogen provide independent prognostic information.[38] Thus, persistent plaque inflammation could partly explain the higher cardiac event rate of cTnT-positive patients. Conjointly with cTnT elevations, we and others have observed an activation of coagulation and platelets as suggested by the elevation of fibrin monomers, prothrombin fragments, fibrinopeptide A, P-selectin, and CD60,[37,40,41] whereas the fibrinolytic

Table 2

Angiographic Findings in 197 Patients with Acute Coronary Syndromes (Unstable Angina and Non-Q-Wave AMI) in Relation to Admission cTnT

	cTnT <0.1 (n = 152)	cTnT >0.1 (n = 45)	*P* Value
3 VD and/or left main disease	57%	80%	0.004
LV ejection fraction <0.50	31%	51%	0.017
TIMI flow <3	24%	40%	0.031
Intraluminal thrombus	5%	13%	0.038
Mean delay to angiography (h)	3.2%	4.7%	<0.01

Data are given as relative frequencies or as mean values with corresponding standard deviation; VD = vessel disease; AMI = acute myocardial infarction; cTnT = cardiac troponin T.

system was not affected.[41] These observations, the angiographic findings, and the results of treatment trials suggest that cTnT elevation may be regarded as a surrogate marker of thrombus formation in unstable lesions leading to micronecrosis due to embolization of microthrombi or exaggerated local vasoconstriction.

To test these hypotheses, we conducted a prospective angiographic evaluation of consecutive patients with acute coronary syndromes admitted within 48 hours of the last chest pain episode, excluding those with ST-segment elevations.[42] Cardiac TnT was analyzed on admission and before angiography. The cTnT-positive patients with unstable angina had a depressed left ventricular function, a higher prevalence of three-vessel or left-main disease, a higher degree of culprit lesion resulting in more severe flow reduction, and a higher prevalence of thrombus formation more often than did cTnT-negative patients (Table 2). The presence of cTnT in unstable angina patients was associated with an 8.5-fold increased risk for the presence of a critical lesion (high-grade stenosis with flow reduction or thrombus formation, or both) and a 3.8-fold increased risk for the presence of severe coronary artery disease (impaired left ventricular function, left main stenosis, three-vessel disease). Concordant angiographic findings regarding culprit lesion morphology and the prevalence of local thrombus have been reported by Heeschen and coworkers.[43] These investigators analyzed angiographic findings of the CAPTURE trial, which recruited patients with unstable angina refractory to medical treatment. They reported on the presence of thrombus in 14.6% versus 4.2%, complex lesion morphology in 72% versus 28%, and TIMI classification flow less than 2 in 40.6% versus

25.5% in cTnT-positive patients as compared with cTnT-negative patients. Moreover, they showed that abciximab therapy reduced visible thrombus in the cTnT-positive patients from 14% to 5.1% (63% reduction; $P<0.001$), whereas there was no documented thrombus resolution among cTnT-negative patients. Thus, it appears that the short-term outcome of cTnT-positive patients with unstable angina is dominated by culprit lesion instability and thrombus formation, whereas long-term outcome may be more dependent on plaque inflammation and the severity of coronary artery disease.

Therapeutic Implications

It is unknown at present whether patients with ST-segment elevation AMI may benefit from guidance of therapy according to their admission cTnT status and, if so, what the optimal adjunctive treatment strategies might be. By contrast, in patients with acute coronary syndrome but without ST-segment elevation, there is growing evidence that cTnT-positive patients benefit from aggressive antithrombotic and antiplatelet therapy, whereas in cTnT-negative patients these therapeutic measures do not improve survival and reduce cardiac events acutely and at 6 months follow-up.

The FRISC I study was the first trial to indicate prospectively that admission cTnT values may be useful for the selection of treatment strategies.[14] In this trial of 971 patients, dalteparin was highly effective in the cTnT-positive patients by reducing the cardiac event rate from 14.2% under placebo to 7.4% with dalteparin at 40 days follow-up. However, in the cTnT-negative patients with unstable angina, the cardiac event rate in the dalteparin and the placebo group was not different (5.7% versus 4.7%). Thus, the dalteparin medication could have been restricted to the cTnT-positive patients without adverse effects on outcome.

The CAPTURE study group analyzed the efficiency of abciximab therapy in patients with angina refractory to medical therapy and subjected to percutaneous interventions.[44] The blood samples were retrospectively analyzed for their cTnT status. A positive admission cTnT identified those who would benefit most from abciximab therapy. At 72 hours after randomization, rates of nonfatal MI and death were significantly lower in these patients (3.6% versus 17.4%; $P=0.007$) as compared with the placebo-treated group. These differences in the cardiac event rates persisted at 6 months follow-up (9.5% versus 23.9%; $P=0.002$). By contrast, cardiac event rates were not significantly different between abciximab and placebo therapy among cTnT-negative patients (4.5% versus 4.2%). Thus, the investigators concluded that cTnT-positive patients with unstable angina should be treated with abciximab if they are scheduled for percutaneous intervention, whereas there is no need to treat cTnT-negative patients.

Recently, the troponin results of the PRISM-PLUS trial were also reported.[45] The PRISM-PLUS trial included 2222 patients with unstable angina treated with tirofiban versus placebo. In the retrospective analysis of troponin in available blood samples, it was observed that active treatment reduced rates of death and MI only in the troponin-positive patients (30-day event rate 3.5% in tirofiban and 13.7% in placebo, 77% risk reduction), whereas 30-day mortality rates were similar in troponin-negative patients in the tirofiban and placebo groups (5.9% versus 4.7%).

In a substudy analysis, the FRISC II investigators tested prospectively the role of early percutaneous interventions in unstable angina patients with different cTnT strata.[46] The treatment effects were analyzed in three cTnT groups (Group A: cTnT <0.01 $\mu g/L$, Group B: cTnT 0.01 to 0.1 $\mu g/L$; Group C: cTnT >0.1 $\mu g/L$). The authors reported lower rates for nonfatal MI and death after early invasive therapy in patients with elevated cTnT (Groups B and C) but not in patients with a cTnT less than 0.01 $\mu g/L$ (Group A). The relative risk reduction of early invasive versus conservative therapy was 29% ($P=0.015$) and 38% ($P=0.005$) in Groups C and B, respectively. Thus, this study also supports the hypothesis that elevated cTnT indicates a high-risk subgroup that benefits from early aggressive treatment.

Conclusion

The cTnT assay has emerged from a stage of scientific evaluation to become a standardized diagnostic tool for centralized or bedside testing. Its analytical sensitivity and precision enable the detection of minor myocardial damage that escapes routine cardiac enzyme measurements. It has been shown in many independent multicenter trials that a cTnT level greater than 0.1 $\mu g/L$ readily identifies patients with an increased rate of cardiac events at short and long term. These findings apply to patients with ST-segment elevation AMI and patients with acute coronary syndromes but without ST-segment elevation. Patients who have ST-segment elevation and who are positive for cTnT on admission represent a group that presents later after the onset of symptoms and has most probably less efficient reperfusion, more severe coronary disease, and poorer left ventricular function. In unstable angina patients, elevated cTnT levels are associated with a higher grade index lesion, more thrombus load, and more severe coronary artery disease. Treatment trials on the unstable angina patients indicate that predominately cTnT-positive patients benefit from aggressive antithrombotic and antiplatelet therapy and early interventions. By contrast, cTnT-negative patients are a low-risk subgroup that does not benefit from these treatment options since the long-term cardiac event rate seems to be determined by mechanisms aside from plaque instability.

In conclusion, cTnT has had an impact on the proper classification of chest pain patients and has put the former gold standard CK-MB in question. It appears that the time of CK-MB-based diagnosis of AMI has passed, and the troponins may play a more definitive role in diagnostic classification, risk stratification, and guidance of therapy in the near future.

References

1. Katus HA, Khaw BA, Mizusawa E, et al. Circulating cardiac myosin light chains in myocardial infarction: Detection by radioimmunoassay. *Circulation* 1979;59(suppl II):539.
2. Katus HA, Hurrell JG, Matsueda GR, et al. Increased specificity in human cardiac myosin radioimmunoassay utilizing two monoclonal antibodies in a double sandwich assay. *Mol Immunol* 1982;19:451–455.
3. Katus HA, Yasuda T, Gold HK, et al. Diagnosis of acute myocardial infarction by detection of circulating cardiac myosin light chains. *Am J Cardiol* 1984;54:964–970.
4. Katus HA, Diederich KW, Uellner M, et al. Myosin light chain release in acute myocardial infarction: Noninvasive estimation of infarct size. *Cardiovasc Res* 1988;22:456.
5. Katus HA, Remppis A, Looser S, et al. Enzyme linked immuno assay of cardiac troponin T for the detection of acute myocardial infarction in patients. *J Mol Cell Cardiol* 1989;21:1349–1353.
6. Katus HA, Remppis A, Neumann FJ, et al. Diagnostic efficiency of troponin T measurements in acute myocardial infarction. *Circulation* 1991;83: 902–912.
7. Katus HA, Looser S, Hallermayer K, et al. Development and in vitro characterization of a new immunoassay of cardiac troponin T. *Clin Chem* 1992; 38:386–393.
8. Katus HA, Remppis A, Scheffold T, et al. Intracellular compartmentation of cardiac troponin T and its release kinetics in patients with reperfused and non-reperfused myocardial infarction. *Am J Cardiol* 1991;67:1360–1367.
9. Katus HA, Schoeppenthau M, Tanzeem A, et al. Non invasive assessment of perioperative myocardial cell damage by circulating cardiac troponin T. *Br Heart J* 1991;65:259–264.
10. Hamm CW, Ravkilde J, Gerhardt W, et al. The prognostic value of serum troponin T in unstable angina. *N Engl J Med* 1992;327:146–150.
11. Antman EM, Tanasijevic MJ, Thompson B, et al. Cardiac-specific troponin I levels predict the risk of mortality in patients with acute coronary syndromes. *N Engl J Med* 1996;335:1342–1349.
12. Lindahl B, Venge P, Wallentin L. Relation between troponin T and the risk

of subsequent cardiac events in unstable coronary artery disease. The FRISC Study Group. *Circulation* 1996;93:1651–1657.
13. Ohman EM, Armstrong PW, Christenson RH, et al. Risk stratification with admission cardiac troponin T levels in acute myocardial infarction. *N Engl J Med* 1996;335:1333–1341.
14. Lindahl B, Venge P, Wallentin L. Troponin T identifies patients with unstable coronary artery disease who benefit from long-term antithrombotic protection. FRISC Study Group. *J Am Coll Cardiol* 1997;29:43–48.
15. Müller-Bardorff M, Freitag H, Scheffold T, et al. Development and characterization of a rapid assay for bedside determinations of cardiac troponin T. *Circulation* 1995;92:2869–2875.
16. Müller-Bardorff M, Hallermayer K, Schröder A, et al. Improved troponin T ELISA specific for cardiac troponin T isoform: Assay development and analytical and clinical validation. *Clin Chem* 1997;43:458–466.
17. Klein G, Kampmann M, Baum H, et al. Clinical performance of the new cardiac markers troponin T and CK-MB on the Elecsys 2010: A multicentre evaluation. *Wien Klin Wochenschr* 1998;3(suppl):40–51.
18. Müller-Bardorff M, Rauscher T, Kampmann M, et al. Development and first evaluation of a device for quantitative bedside testing of cardiac troponin T. *Clin Chem* 1999;45:1002–1008.
19. Klein G, Baum H, Gurr E, et al. Multicentre evaluation of two new assays for myoglobin and troponin T on the Elecsys 2010 and 1010 analysers. *Clin Chem* 1999;45:495A.
20. Collinson PO, Gerhardt W, Katus HA, et al. Multicenter study of a rapid troponin T test with improved detectability: Analytical and clinical performance in acute and subacute myocardial infarction. *Eur J Clin Chem* 1996; 34:591–598.
21. Gerhardt W, Ljungdahl L, Collinson PO, et al. An improved rapid troponin T test with a decreased detection limit: A multicentre study of the analytical and clinical performance in suspected myocardial damage. *Scand J Clin Lab Invest* 1997;57:549–558.
22. Hirschl MM, Herkner H, Laggner AN, et al. Analytical and clinical performance of an improved qualitative troponin T rapid test in laboratories and critical care units. *Arch Pathol Lab Med* 2000;124:583–587.
23. Hamm CW, Goldmann BU, Heeschen C, et al. Emergency room triage of patients with acute chest pain by means of rapid testing for cardiac troponin T or troponin I. *N Engl J Med* 1997;337:1648–1653.
24. Lindahl B, Diderholm E, Venge P, et al. Comparison of qualitative versus quantitative determination of troponin T in unstable coronary artery disease. *Eur Heart J* 1999;20(suppl):P2105.
25. Gerhardt W, Nordin G, Ljungdahl L. Can troponin T replace CKMB mass as gold standard for acute myocardial infarction. *Scand J Clin Lab Invest* 1999;59(suppl 230):83–89.
26. Zimmermann R, Baki S, Dengler TJ, et al. Troponin T release after heart transplantation. *Br Heart J* 1993;69:395–398.
27. Dengler TJ, Zimmermann R, Braun K, et al. Elevated serum concentrations of cardiac troponin T in acute allograft rejection after human heart transplantation. *J Am Coll Cardiol* 1998;32:405–412.
28. Bachmaier K, Mair J, Offner F, et al. Serum cardiac troponin T and creatine kinase-MB elevations in murine autoimmune myocarditis. *Circulation* 1995; 92:1927–1932.
29. Ferjani M, Droc G, Dreux S, et al. Circulating cardiac troponin T in myocardial contusion. *Chest* 1997;111:427–433.

30. Müller-Bardorff M, Giannitsis E, Weidtmann B, et al. Prognostic value of troponin T in patients with pulmonary embolism. *Circulation* 1999; 100(suppl I):3102A. Abstract.
31. Stubbs P, Collinson P, Moseley D, et al. The prognostic significance of admission troponin T concentrations in myocardial infarction. *Circulation* 1996;94:1291–1297.
32. Ohman EM. GUSTO-III Symposium. XIXth Congress of the European Society of Cardiology. Stockholm; 1997.
33. Frostfeldt G, Lindahl B, Nygren A, et al. Possible reasons for the prognostic value of troponin T on admission in patients with ST-segment elevation myocardial infarction. *Eur Heart J* 1999;20(suppl):P2109. Abstract.
34. Ramanathan K, Stewart T. Admission troponin T level may predict 90 minute TIMI flow after thrombolysis. *Circulation* 1997;96(suppl):I270. Abstract.
35. Kurowski V, Killermann D, Frey N, et al. A positive troponin T-test on admission independently predicts an adverse prognosis in patients with acute myocardial infarction and direct PTCA. *Circulation* 1999;100(suppl): I373. Abstract.
36. Ravkilde J, Nissen H, Horder M, et al. Independent prognostic value of serum creatine kinase isoenzyme MB mass, cardiac troponin T and myosin light chain levels in suspected acute myocardial infarction. *J Am Coll Cardiol* 1995;25:574–581.
37. Davies MJ, Thomas AC, Knapman PA, et al. Intramyocardial platelet aggregation in patients with unstable angina suffering sudden ischemic cardiac death. *Circulation* 1986;73:418–427.
38. Liuzzo G, Biasucci LM, Gallimore JR, et al. Prognostic value of C-reactive protein and serum amyloid A in severe unstable angina. *N Engl J Med* 1994; 331:417–424.
39. Haverkate F, Thompson SG, Pyke SDM, et al. Production of C-reactive protein and risk of coronary events in stable and unstable angina. *Lancet* 1997;349:462–466.
40. Terres W, Kummel P, Sudrow A, et al. Enhanced coagulation activation in troponin T-positive unstable angina pectoris. *Am Heart J* 1998;135:281–286.
41. Giannitsis E, Müller-Bardorff M, Schweikart S, et al. Relationship of troponin T and procoagulant activity in unstable angina. *Haemost Thromb* 2000; 83:224–228.
42. Frey N, Dietz R, Tolg R, et al. Quantitative Koronarangiographie bei Patienten mit instabiler Angina pectoris in Abhängigkeit vom Ergebnis des Troponin T Tests. *Zeitschr Kardiol* 1999;88(suppl 1):P507. Abstract.
43. Heeschen C, van der Brand MJ, Hamm CW, et al, for the CAPTURE Investigators. Angiographic findings in patients with refractory unstable angina according to troponin T status. *Circulation* 1999;104:1509–1514.
44. Hamm CW, Heeschen C, Goldmann B, et al, for the CAPTURE Study Investigators. Benefit of ABCIXIMAB in patients with refractory unstable angina in relation to serum troponin T levels. *N Engl J Med* 1999;340:1623–1629.
45. Heeschen C, Hamm CW, Goldmann B, et al, for the PRISM Study Investigators. Troponin concentrations for stratification of patients with acute coronary syndromes in relation to therapeutic efficacy of tirofiban. *Lancet* 1999; 354:1757–1762.
46. Lagerqvist B, Diderholm E, Lindahl B, et al. An early invasive treatment strategy reduces cardiac events regardless of troponin levels in unstable coronary artery disease with and without troponin-elevation: A FRISC-II substudy. *Circulation* 1999;100(suppl):I497.

Chapter 12

Creatine Kinase:

A Marker for the Early Diagnosis of Acute Myocardial Infarction

Robert Fromm, MD, MPH and Robert Roberts, MD

Introduction

Several million patients present annually with chest pain syndromes.[1] These patients seek care in emergency departments and physicians' offices, and many are admitted to the hospital for further evaluation. Ultimately, only about 10% are subsequently proven to have myocardial infarction (MI). This observation underscores the fact that symptoms and signs do not differentiate between MI and ischemia in most patients, and noncardiac causes of chest pain are frequently not evident on initial evaluation. The initial electrocardiogram (ECG) is only of limited utility, with diagnostic changes in only about 40% of patients subsequently shown to have infarction; this represents less than 5% of all patients presenting with chest pain.[2,3]

Traditionally, the diagnosis of MI has been determined from the history, physical examination, ECG, and the evaluation of serum markers of myocardial necrosis. The classic serum marker has been myocardial muscle creatine kinase isoenzyme (CK-MB).[2,3] Unfortunately, conventional total CK-MB is not reliable in excluding infarction until at least 10 to 12 hours after the onset of chest pain.[4] Advances in the therapy of MI have led to increased interest in improving the diagnosis. Fibrinolysis within the first hour of onset results in a 90% reduction in mortality compared with little to no reduction in mortality if delayed 10 to 12 hours.[5] In contrast, fibrinolytic therapy administered to patients with unstable angina leads to increased death and infarction.[6] Recently,

From: Adams JE III, Apple FS, Jaffe AS, Wu AHB (eds). *Markers in Cardiology: Current and Future Clinical Applications.* Armonk, NY: Futura Publishing Company, Inc.; © 2001.

new markers of cardiac injury, cardiac troponin T[7] and cardiac troponin I,[8] and easily and rapidly performed quantitative assays for myoglobin[9] and CK-MB subforms[4] have been developed with the goal of more rapid and specific diagnosis of infarction. CK-MB subforms have proven themselves to be particularly useful in the early diagnosis of MI.

In the initial hours after the onset of MI, the amount of CK-MB released from the myocardium is quite small and after dilution in the circulation blood volume, generally does not exceed the upper limit of the normal range.[10,11] Although there is only one form[12] of CK-MB in myocardial tissue (CK-MB_2), on its release into the blood, lysine carboxypeptidase cleaves the positively charged terminal lysine from the M subunit,[13] producing a more negatively charged molecule (CK-MB_1). Normally, CK-MB_2 and CK-MB_1 are in equilibrium, and the absolute levels may be as low as 0.5 to 1 U/L.[10,11] Subforms of CK-MB can be quantified using high-voltage electrophoresis, and the ratio of CK-MB_2 to CK-MB_1 can be determined. This ratio of tissue to circulating CK-MB has been successfully used in the early diagnosis of acute MI (AMI).

We recently reported the performance of CK-MB subforms in a large cohort of chest pain patients in the Diagnostic Marker Cooperative Study (DMCS).[14] This multicenter, prospective, double-blind study involved Baylor College of Medicine and the University of Texas Medical School. Enrollment was consecutive, 24 hours per day, 7 days per week, at four affiliated teaching hospitals in Houston, Texas: Ben Taub General Hospital, Hermann Hospital, The Methodist Hospital, and Veterans Affairs Medical Center. To be eligible for enrollment, patients had to be 21 years of age with chest pain ≥15 minutes suspected to be myocardial in origin and occurring within 24 hours of presentation.

Serial blood samples were obtained on arrival, 1 hour after arrival, then every 2 hours up to 6 hours from the onset of chest pain, and every 4 hours thereafter for 24 hours. Samples were placed on ice for transport to the core laboratory for analysis. The management and disposition of the patient were determined by the attending physician using standard practices. Attending physicians did not have access to results of the cardiac markers performed by the Core Laboratory.

CK-MB Subform Assay

Blood for CK-MB subforms was collected in ethylenediaminetetraacetic acid (EDTA),[4] and the plasma was recovered. Standard controls were assayed routinely. Total CK activity was quantitated using a spectrophotometric enzymatic assay (Sigma Diagnostics, St. Louis, MO) with an upper normal limit of 120 IU/L. Total plasma CK-MB mass was measured with the Stratus CK-MB Fluorometric Enzyme Immunoassay (Dade Intentional Inc., Miami, FL) with an upper normal limit of 7 μg/L. CK-MB subforms (MB_1 and MB_2) and total serum CK-MB activity were quantitated using the Cardio REP CK Isoforms Procedure (Helena Laboratories, Beaumont, TX), with upper limits for MB_2 of 2.5 IU/L and a ratio of MB_2/MB_1 of 1.6.

Criteria for the Diagnosis of Infarction

The diagnostic standard for MI was a CK-MB mass $\geq$7 μg/L and a CK-MB index (CK-MB mass/CK) $\geq$2.5%, determined by the results of the Core Laboratory in two or more samples obtained in the first 24 hours after hospital arrival or in one sample if only one sample were available for analysis. Thus the current gold standard marker was the criteria upon which CK-MB subforms were assessed. Infarction was further characterized by electrocardiographic changes into Q-wave and non-Q-wave infarcts. As would be expected from a large population of chest-pain patients, time from chest pain onset to presentation varied with a mean time of presentation of 5.5 ± 4.7 hours.

Results

There were 955 patients enrolled, and MI was confirmed by CK-MB mass criteria in 119 (12.5%). The ECG was diagnostic (ST-segment elevation) in only 45% of infarct patients. There were 36 (30.3%) patients who had Q-wave infarcts, and the remaining 83 had non-Q-wave infarction. In the 119 patients with infarction, reperfusion therapy was administered to 34.4%: thrombolytic therapy in 30 patients (25.2%) and primary percutaneous transluminal coronary angioplasty in 11 (9.2%). Unstable angina was diagnosed in 203 (21.3%) patients, and the remaining patients were diagnosed with other conditions. The performance of CK-MB subforms in the population is depicted in Table 1. The sensitivity and specificity of CK-MB subforms based on the first sample were 48.7% and 87.6%, respectively. The sensitivity and specificity of CK-MB subforms based on the combined first and second samples were 81.6% and 84.4%, respectively.

The results of this study confirm the utility of CK-MB subforms in the diagnosis of AMI. Other studies have consistently shown the CK-MB subforms to be the most sensitive and specific marker within 6 hours of the onset of symptoms.[15] The negative predictive value of CK-MB subforms at 6 hours from the onset of chest pain was 97%

Table 1

Diagnostic Sensitivity and Specificity of Markers for Myocardial Infarction Based on Time from Onset of Chest Pain

	Early Diagnosis			Late Diagnosis			
Time (Hours)	2	4	6	10	14	18	22
Marker:							
CK-MB Subforms							
Sensitivity (%)	21.1	46.4	91.5	96.2	90.6	80.9	53.1
Specificity (%)	90.5	88.9	89.0	90.2	90.0	89.9	92.2
Troponin T							
Sensitivity (%)	10.5	35.7	61.7	86.5	84.9	78.7	85.7
Specificity (%)	98.4	98.3	96.1	96.4	96.1	95.7	94.6
Troponin I							
Sensitivity (%)	15.8	35.7	57.5	92.3	90.6	95.7	89.8
Specificity (%)	96.8	94.2	94.3	94.6	92.2	93.4	94.2

CK-MB = MB isoenzyme of creatine kinase.

in the DMCS, an important observation when one considers that the probability of infarction in this population is only ~10%. Thus, a negative CK-MB subform analysis at 6 hours after the onset of chest pain identifies a very low-risk population and when combined with clinical assessment may permit the triage of low-risk patients to less expensive and less resource-consuming management strategies. We[4] have shown that early triaging based on CK-MB subforms could potentially save billions of dollars by avoiding unnecessary hospital admissions. In a pilot study performed in 1314 patients with chest pain, admission to the hospital on the basis of CK-MB subforms reduced the cost per patient by 35%. These results applied nationally would represent billions of dollars saved.[16]

The DMCS also compared the performance of CK-MB subforms with the other available serum markers of myocardial injury, namely, cardiac troponin T and I, myoglobin, and total CK-MB. CK-MB subforms were the most reliable marker for the early (within 6 hours of the onset of chest pain) diagnosis of MI, and the total CK-MB activity, derived from the CK-MB subforms, was the most sensitive marker for late diagnosis. The area under the receiver operating characteristic curve, an integrated index of diagnostic test performance, was 0.950 at 6 hours after the onset of chest pain. This study confirmed the high specificity but late rise in sensitivity of the cardiac troponin for the diagnosis of MI. The assay for CK-MB subforms is an automated process that requires approximately 25 minutes for completion. The costs and time requirements are similar to those of other diagnostic markers of myocardial injury.

Results in Relation to Other Studies

Morris and colleagues[17] also examined CK-MB subforms and troponin I in a cohort of emergency department patients presenting within 12 hours of the onset of chest pain. Their study also demonstrated that CK-MB subforms had the highest sensitivity for the diagnosis of MI. Until recently, there has been no large systematic study comparing the biochemical markers for early diagnosis of infarction. A recent study involved 385 patients presenting to the emergency room within 6 hours of the onset of chest pain who underwent testing for CK-MB subforms and myoglobin.[18] There were 50 patients with confirmed AMI, and the subforms based on samples analyzed within the first 6 hours provided 96% sensitivity and 82% specificity. If the criterion of total CK-MB greater than 4 U/L was added to the diagnostic criteria, CK-MB subforms exhibited a sensitivity of 98% and a specificity of 86%. In contrast, the myoglobin within the 6-hour interval had a sensitivity of 80% and a specificity of 93%. The negative predictability value for the subforms at 6 hours from the onset of symptoms was 98%. This 98% value indicates that if CK-MB subform is normal at 6 hours, there is less than a 2% chance that the patient has infarction. It is very important to realize the reliability of a negative test in this situation since less than 10% of patients with chest pain will have an AMI. In addition to a false-negative diagnosis leading to the lack of appropriate therapy, there are also the legal implications.

The results of the DMCS[14] and of the study by McEvoy and colleagues[19] confirm CK-MB subforms as the most sensitive and specific marker for the early diagnosis of infarction compared with myoglobin, total CK-MB, and the cardiac troponins. Routine use of CK-MB subforms in the emergency room for patients with chest pain should provide for more appropriate patient therapy, more rapid administration, fewer unnecessary hospital admissions, and reduced cost. This will be particularly important in patients with non-Q-wave infarction who present without ST-segment elevation, which is the most common type of AMI.[20,21] Since thrombolytic therapy is contraindicated in those patients with normal CK-MB but antiplatelet therapy is appropriate, CK-MB could provide for more appropriate and rapid therapeutic stratification.[22]

References

1. Selker HP. Coronary care unit triage decision aids: How do we know when they work? *Am J Med* 1989;87:491–493.
2. Fisch C. The clinical electrocardiogram: Sensitivity and specificity. In Fisch C (ed): *ACC Current Journal Review.* New York: Elsevier Science, Inc.; 1997: 71–75.
3. Lee TH, Rouan GW, Weisberg MC, et al. Sensitivity of routine clinical criteria for diagnosing myocardial infarction within 24 hours of hospitalization. *Ann Intern Med* 1987;106:181–186.
4. Puleo PR, Meyer D, Wathen C, et al. Use of rapid assay of subforms of creatine kinase MB to diagnose or rule out acute myocardial infarction. *N Engl J Med* 1994;331:561–566.
5. Weaver WD, Cerqueira M, Halstrom AP, et al. Prehospital-initiated vs hospital-initiated thrombolytic therapy. The Myocardial Infarction Triage and Intervention Trial. *JAMA* 1993;270:1211–1216.
6. TIMI IIIB Investigators. Effects of tissue plasminogen activator and a comparison of early invasive and conservative strategies in unstable angina and non-Q-wave myocardial infarction: Results of the TIMI III B Trial. *Circulation* 1994;89:1545–1556.
7. Katus HA, Remppis A, Neumann FJ, et al. Diagnostic efficiency of troponin T measurements in acute myocardial infarction. *Circulation* 1991;83: 902–912.
8. Adams JE III, Schechtman KB, Landt Y, et al. Comparable detection of acute myocardial infarction by creatine kinase MB isoenzyme and cardiac troponin I. *Clin Chem* 1994;40:1291–1295.
9. Montague C, Kircher T. Myoglobin in the early evaluation of acute chest pain. *Am J Clin Pathol* 1995;104:472–476.
10. Puleo PR, Guadagno PA, Roberts R, Perryman MB. Sensitive, rapid assay of subforms of creatine kinase MB in plasma. *Clin Chem* 1989;35:1452–1455.
11. Puleo PR, Guadagno PA, Roberts R, et al. Early diagnosis of acute myocardial infarction based on assay for subforms of creatine kinase-MB. *Circulation* 1990;82:759–764.
12. Wevers RA, Delsing M, Klein-Gebbink JA, Soons JB. Post-synthetic changes in creatine kinase isoenzymes. *Clin Chim Acta* 1978;86:323–327.
13. George S, Ishikawa Y, Perryman MB, Roberts R. Purification and characterization of naturally occurring and in vitro induced multiple forms of MM creatine kinase. *J Biol Chem* 1984;259:2667–2674.
14. Zimmerman J, Fromm R, Meyer D, et al. Diagnostic Marker Cooperative Study (DMCS) for the diagnosis of myocardial infarction. *Circulation* 1999; 99:1671–1677.
15. Bernardoni C, Pravettoni G, Besomi G, et al. Rapid assay of isoforms of creatine kinase MB to diagnose acute myocardial infarction (AMI). *J Suisse de Medicine* 1995;29:32S. Abstract.
16. Trahey TF, Dunevant SL, Thompson AB, et al. Early hospital discharge of chest pain patients using creatine kinase isoforms and stress testing: A community hospital experience. *Circulation* 1996;94:I569.
17. Morris S, Melilli L, Pesce M, et al. Single determinations of troponin I and CKMB isoforms used alone or in combination for early rule-out of acute myocardial infarction. *J Am Coll Cardiol* 1998;31:74A. Abstract.
18. Lustig V, Vandersluis R, Bhargava RK, et al. Clinical evaluation of the CK-

MB isoforms as a marker for acute myocardial infarction. *Clin Chem* 1998; 44:A131. Abstract.

19. McEvoy KB, MacRae AR, Vandersluis R, et al. Optimizing sensitivity and specificity of the Helena Cardio-REP using non-boolean logic. *Clin Biochem* 1998;31:296.
20. Goldberg RJ, Gore JM, Gurwitz JH, et al. The impact of age on the incidence and prognosis of initial acute myocardial infarction: The Worcester Heart Attack Study. *Am Heart J* 1989;117:543–549.
21. Guadagnoli E, Hauptman PJ, Ayanian JZ, et al. Variation in the use of cardiac procedures after acute myocardial infarction. *N Engl J Med* 1995; 333:573–578.
22. Anderson HV, Cannon CP, Williams DO, et al, and TIMI-IIIB Investigators. One-year results of the thrombolysis in myocardial infarction TIMI-IIIB clinical trial. *Circulation* 1994;90:I231. Abstract.

Chapter 13

Fatty Acid Binding Protein as a Plasma Marker for the Early Assessment of Individuals with Acute Coronary Syndromes

Jan F.C. Glatz, PhD and Wim T. Hermens, PhD

Introduction

Biochemical markers of myocardial injury are universally accepted as important determinants for the diagnosis of patients with suspected acute myocardial infarction (AMI), especially in those cases in which electrocardiographic (ECG) changes are equivocal or absent.[1,2] In the last decade, interest in these biochemical markers has increased for two reasons. First, the introduction of new therapeutic strategies has called for earlier and more appropriate diagnosis of patients admitted to the emergency room with chest pain, so as to begin the proper therapy as early as possible. Second, several new plasma markers have been introduced, and some (eg, troponin T [TnT]) allow for the assessment of patients with not only MI but also unstable angina and prolonged chest pains. Some markers even provide prognostic value.[3]

Fatty acid binding protein (FABP) has similarly been proposed as an early plasma marker of acute coronary syndromes. In this chapter, we describe some relevant features of this protein and summarize the studies that have investigated its application as a marker of myocardial injury. The release and plasma kinetics of FABP closely resemble those of myoglobin, but the relatively low plasma reference concentration of

Work in the authors' laboratory was supported by grants from the Netherlands Heart Foundation (D90.003, 95.189, and 98.063), the Ministry of Economic Affairs (StiPT/MTR 88.002 and BTS 97.188), and the European Community (BMH1-CT93.1692 and CIPD-CT94.0273).

From: Adams JE III, Apple FS, Jaffe AS, Wu AHB (eds). *Markers in Cardiology: Current and Future Clinical Applications.* Armonk, NY: Futura Publishing Company, Inc.; © 2001.

FABP makes it superior to myoglobin for the monitoring of myocardial injury.

Biochemistry and Biological Function of FABP

In the soluble cytoplasm of almost all tissue cells, a relatively small (14 to 15 kd) protein is found that can reversibly and noncovalently bind long-chain fatty acids and is therefore called (cytoplasmic) FABP. At least nine distinct types of FABP occur, and these are generally named after the tissue in which they were first identified and/or mainly occur, eg, H (heart)-FABP, L (liver)-FABP, and I (intestinal)-FABP. Subsequent studies revealed the presence of L-FABP also in small intestine and the presence of H-FABP also in skeletal muscle, in distal tubule cells of the kidney, and in some parts of the brain.[4] The FABPs are relatively abundant in tissues with an active fatty acid metabolism such as liver, adipose tissue, and heart, in which each shows a tissue content of 0.5 to 1 mg FABP per g wet weight of tissue.[4]

The FABPs belong to a multigene family of intracellular lipid-binding proteins that also includes the cellular retinoid-binding proteins.[4,5] These proteins each contain 126 to 137 amino acid residues (molecular mass 14 to 15 kd) and show a similar tertiary structure which resembles that of a clam shell (Fig. 1).[5] The lipid ligand is bound in between the two halves of the clam by interaction with specific amino acid residues within the binding pocket, a so-called β-barrel, of the protein. Human H-FABP contains 132 amino acid residues (14.5 kd) and is an acidic protein (p*I* 5).[6,7]

The FABPs appear to be stable proteins, exhibiting an intracellular turnover with a half-life of approximately 2 to 3 days.[4] Their cellular expression is regulated primarily at the transcriptional level. In general, the FABP expression is responsive to changes in lipid metabolic activity as induced by various (patho)physiological and pharmacologic stimuli.[4] For instance, the H-FABP content of heart and skeletal muscles increases by endurance training[8] and is also higher in the diabetic state,[9] but is slightly decreased in the hypertrophied heart.[10,11]

The primary biological function of the FABPs is their facilitation of the cytoplasmic translocation of long-chain fatty acids, which is normally hampered by the very low solubility of these compounds in aqueous solutions. Therefore, FABP can be regarded as an intracellular counterpart of plasma albumin. Definite proof of this function was obtained recently, when it was found that cardiac myocytes isolated from mice lacking the H-FABP gene showed a markedly lower (approximately −50%) rate of fatty acid uptake and oxidation.[12] Other postulated functions for H-FABP include a participation in signal transduction pathways such as fatty acid regulation of gene expression,[13,14] and

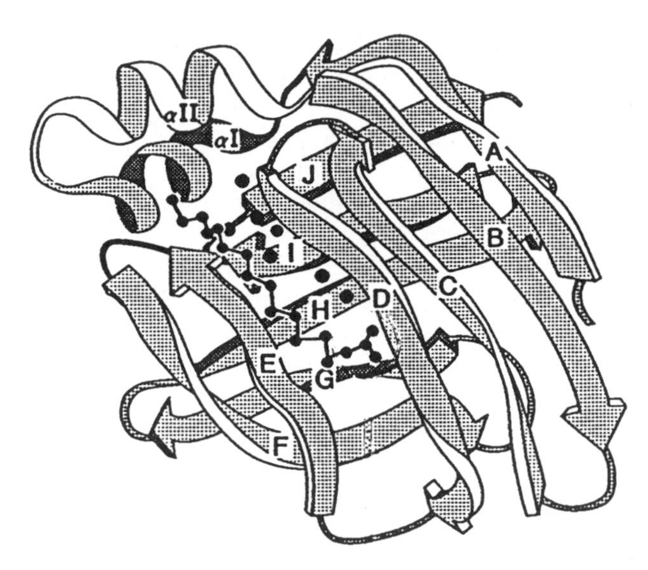

Figure 1. A schematic view of fatty acid binding protein (FABP) with bound long-chain fatty acid. A ribbon diagram is shown for I (intestinal-type)-FABP, highlighting the 10 β-strands (βA-βJ) forming two so-called β-sheets in between which the fatty acid ligand is bound, and two short α-helical domains (αI and αII). Human H (heart type)-FABP has a similar structure, but the fatty acid is bound in a U-shaped conformation rather than in an elongated form.[73] Reproduced from Reference 74, with permission.

the protection of myocytes against the adverse (detergent-like) effects of long-chain fatty acids.[4] The latter function would be of special importance for the ischemic heart because the tissue accumulation of fatty acids and their derivatives occurring in this condition has been associated with arrhythmias, increased myocardial infarct size, and depressed myocardial contractility.[15] The presence of H-FABP then may be crucial to sequester accumulating fatty acids and thus prevent tissue injury. However, current, available evidence for such a role for H-FABP remains inconclusive.[4]

Immunochemical Assay of FABP in Plasma

Because FABP is a nonenzymatic protein, its detection and quantification must be performed with an immunochemical assay. A large number of immunoassays for H-FABP have been described, mostly enzyme-linked immunosorbent assays (ELISAs) of the antigen capture type,[16–19] but also a competitive immunoassay[20] and an immunofluorometric assay.[21] In most cases, monoclonal antibodies are used, and these show virtually no cross-reactivity with other FABP types.[7,22] Recombinant H-FABP appears immunochemically equivalent to the tissue-derived protein and, therefore, is now commonly applied as standard in the immunoassay.[7,16]

These assays have been used successfully for retrospective analyses of plasma FABP in patient samples. However, the implication of these tests for clinical decision making in the case of suspected AMI is hampered by the fact that the reported fastest immunoassay[16] still takes 45 minutes to complete. Therefore, more rapid FABP immunoassays are being developed. To date, these include a microparticle-enhanced turbidimetric assay to be performed on a conventional clinical chemistry analyzer (performance time 10 minutes),[23] an automated sandwich immunoassay (performance time 23 minutes),[24] and an electrochemical immunosensor (performance time 20 minutes).[25,26] The electrochemical immunosensor is based on screen-printed graphite electrodes and uses an immunosandwich procedure and an amperometric detection system.[26] Measurements of plasma samples from patients with AMI with this immunosensor and with an ELISA show an excellent correlation.[26,27] More recently, a new principle for rapid immunoassay of proteins based on in situ precipitate-enhanced ellipsometry was presented and applied for assay of FABP.[28] This technique enables the development of a one-step ELISA with a performance time of less than 10 minutes. With the exception of the ELISA assays, calibrated with recombinant FABP, these new techniques still require further evaluation and standardization.

Release and Elimination of FABP upon Muscle Injury

The release of H-FABP from injured muscle was first demonstrated in 1988 in isolated working rat hearts,[29] and indicated the potential use of FABP as a plasma marker of myocardial injury in humans. Subsequently, several groups reported the release of FABP into plasma of patients with AMI.[17,18,20,30] The characteristics of the release of FABP from injured myocardium closely resemble those of myoglobin. As an example, Figure 2 shows mean plasma release curves of FABP and of several other plasma marker proteins for 15 AMI patients (treated with thrombolytic therapy) from whom blood samples were obtained frequently during the first 72 hours of hospitalization.[31] In patients treated with standard thrombolytic therapy after AMI, peak plasma concentrations of FABP and myoglobin are reached after approximately 4 hours after first symptoms, whereas for creatine kinase (CK) this takes approximately 12 hours and for lactate dehydrogenase (LDH) approximately 20 hours. Furthermore, plasma FABP and myoglobin return to their respective reference values within 24 hours after AMI (Fig. 2),

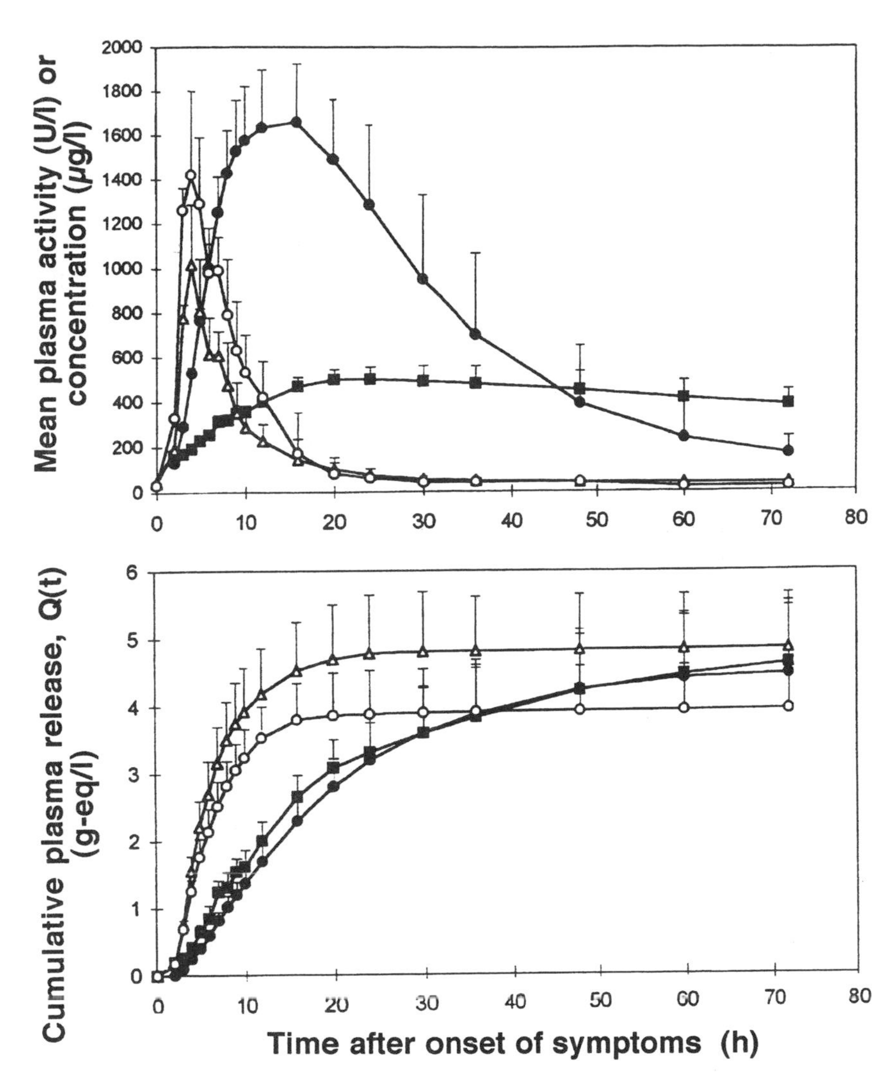

Figure 2. Mean plasma concentration or activities (top panel) and mean cumulative release expressed in gram-equivalents (g-eq) of healthy myocardium per liter of plasma (lower panel) of FABP (multiplied by 10, ○), myoglobin (△), creatine kinase (●), and lactate dehydrogenase isoenzyme-1 (■) as function of time after onset of symptoms in 15 patients after acute myocardial infarction. Data refer to mean ± SEM. Adapted from Reference 31.

indicating the usefulness of both markers particularly for the assessment of a recurrent infarction.[32] However, for AMI patients not treated with thrombolytics, peak levels are reached approximately 8 hours after AMI, and elevated plasma FABP and myoglobin concentrations are found up to 24 to 36 hours after the onset of chest pain.[32] The release of the myofibrillar proteins TnT and troponin I (TnI) from injured myocardium follows a different pattern with elevated plasma concentrations occurring from approximately 8 hours up to more than 1 week after infarction.[3,33] Hence, the so-called diagnostic window of the various marker proteins differs considerably.

The marked differences in the time course of plasma concentrations or activities among the cytoplasmic proteins (CK-MB, LDH, FABP, and myoglobin) are caused by 1) a more rapid washout of the smaller proteins (FABP, myoglobin) from the interstitium to the vascular compartment (Fig. 3), and 2) differences among the proteins in their rate of elimination from plasma.[34] Studies of isolated cardiac myocytes subjected to simulated ischemia showed that protein release from

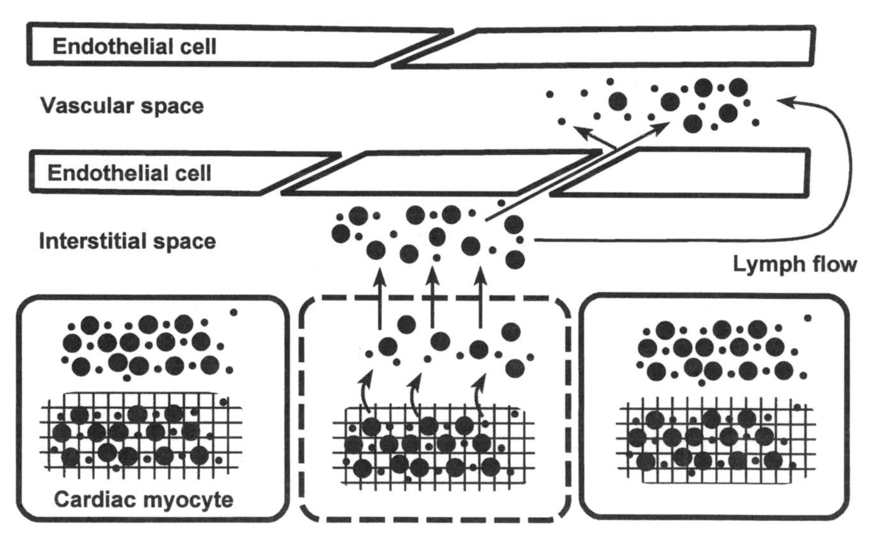

Figure 3. Schematic presentation of the possible transport routes of protein released from damaged cardiac myocytes to the plasma compartment. Proteins can either directly cross the endothelial cell barrier (predominant route for small proteins such as fatty acid binding protein and myoglobin) or they can be transported through lymph drainage (predominant route for larger proteins such as creatine kinase MB and lactate dehydrogenase). Structurally bound proteins (such as troponin T) first must be dissociated from the myofibrillar structures before they can be released into the interstitial space. From Reference 75, with permission.

Table 1

Comparison of Relevant Characteristics for FABP and Myoglobin as Markers of Myocardial Injury

Protein	Molecular Mass (kD)	Cardiac Muscle Content (mg/g)	Skeletal Muscle Content* (mg/g)	Reference Plasma Concentration (μg/L)
FABP	14.5	0.57	0.04–0.14	1.8
Myoglobin	17.6	2.7	2.2–6.7	34

Data are obtained from References 11, 31, 32, and 38. * Range given for muscles of different fiber type composition. FABP = fatty acid binding protein.

the damaged myocytes is independent of molecular mass.[35] This finding indicates that during the protein release phase the sarcolemma does not act as a selective sieve through which small proteins are preferentially lost. The fact that smaller proteins can be detected in blood plasma earlier after muscle injury than can larger proteins therefore relates to a greater permeability of the endothelial barrier for smaller proteins (Fig. 3).[34]

With respect to protein elimination from plasma, FABP and myoglobin, unlike the larger cardiac enzymes, are removed from the circulation predominantly by renal clearance.[17,18,30,34] This explains not only their rapid clearance from plasma after AMI (Fig. 2) but also the maintenance of relatively low plasma concentrations of these proteins in healthy individuals. The plasma concentrations in healthy individuals are determined mainly by the release of protein from skeletal muscle because its total mass far exceeds that of cardiac muscle. Because the skeletal muscle FABP content is relatively low compared with that of myoglobin, the plasma reference concentration of FABP also is relatively low (Table 1).[16,36–39] This notion is also reflected in the ratio of the concentrations of myoglobin and FABP in plasma from healthy subjects (myoglobin:FABP ratio approximately 20),[38] which resembles the ratio in which these proteins occur in skeletal muscle (myoglobin: FABP ratio 15 to 70) (see below).

The role of the kidney in the clearance of FABP and myoglobin from plasma further indicates that increased plasma concentrations of these proteins are likely to be found in case of renal insufficiency. Indeed, it has been reported that patients with chronic renal failure and normal heart function show several-fold increased plasma concentrations of both FABP and myoglobin.[40] In addition, Kleine et al.[18] reported a patient with AMI and severe renal insufficiency in whom the plasma FABP concentration remained markedly elevated for at least 25 hours after infarction.

Discrimination of Cardiac from Skeletal Muscle Injury

A potential drawback of the use of FABP as a plasma marker for monitoring myocardial injury is its presence in significant quantities not only in heart muscle but also in skeletal muscle cells (Table 1). A proper diagnosis of AMI thus may be hampered in the case of extensive skeletal muscle injury such as multiorgan failure, postoperative states, or vigorous exercise. However, this problem can be overcome by the combined measurement of myoglobin and FABP concentrations in plasma and by expressing the ratio of these, because this plasma ratio is a reflection of the ratio in which these proteins occur in the affected tissue cells and it differs between heart muscle (myoglobin:FABP ratio 4 to 5) and skeletal muscles (myoglobin:FABP ratio 20 to 70, depending on type of muscle) (Table 1).[3,32,39]

This finding is illustrated in Figure 4 for patients after AMI in whom the plasma myoglobin:FABP ratio was approximately 5 during the entire period of elevated plasma concentrations (upper panels), and for patients who underwent aortic surgery, which causes no-flow ischemia of the lower extremities, in whom the plasma myoglobin: FABP ratio was approximately 45 (lower panels). In addition, Van Nieuwenhoven et al.[32] described a patient who was defibrillated shortly after AMI, a treatment that most likely results in injury of intercostal pectoral muscles, and in whom the plasma myoglobin:FABP ratio increased from 8 to 60 during the first 24 hours after AMI. Finally, in case AMI patients show a second increase of plasma concentrations of marker proteins, the ratio may be of help to delineate whether this second increase was caused either by a recurrent infarction or by the occurrence of additional skeletal muscle injury. In the former case, the ratio will remain unchanged.[32]

Early Diagnosis of AMI

The application of FABP especially for the early diagnosis of acute coronary syndromes is already indicated from 1) its rapid release into plasma after myocardial injury, and 2) its relatively low plasma reference concentration. Several studies have now firmly established that FABP is an excellent plasma marker for the early differentiation of patients with and those without AMI, and that it even performs better than myoglobin. A selection of these studies is discussed here.

Retrospective analyses of various marker proteins in plasma samples from patients with AMI revealed that the diagnostic sensitivity

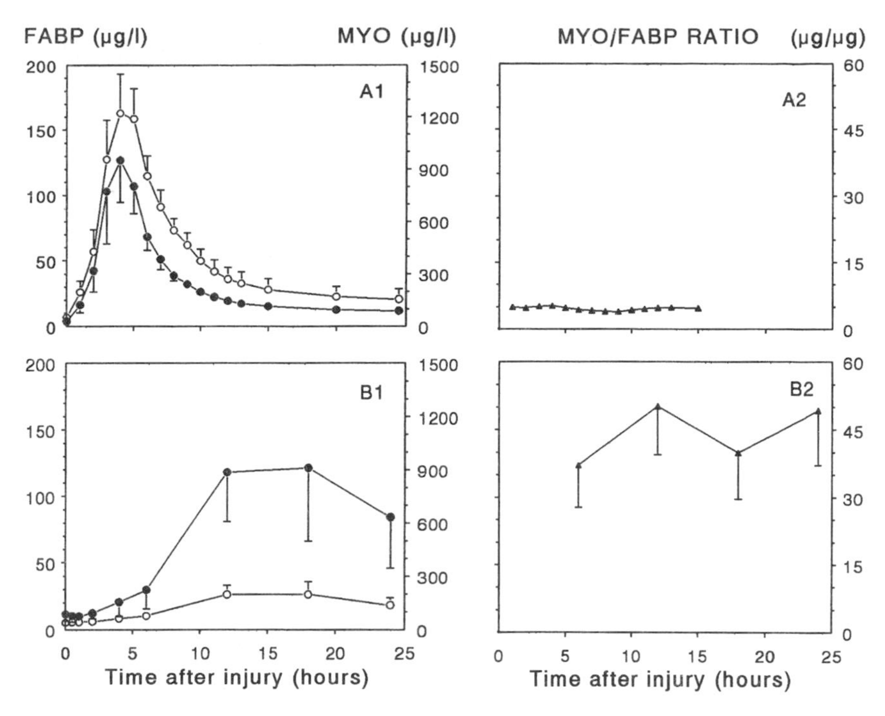

Figure 4. Mean plasma concentrations of myoglobin (MYO; ●) and fatty acid binding protein (FABP; ○) (left panels) and the myoglobin:FABP ratio (▲) (right panels) in nine patients after acute myocardial infarction (and receiving thrombolytic therapy) (A) and in nine patients after aortic surgery (B). Data refer to mean ± SEM. Adapted from Reference 32.

for detection of AMI is better for FABP than for myoglobin or CK-MB, especially in the early hours after the onset of symptoms. For example, in a study including blood samples from 83 patients with confirmed AMI taken immediately upon admission to the hospital (<6 hours after chest pain onset), the diagnostic sensitivity was significantly greater for FABP (78%, CI 67% to 87%) than for myoglobin (53%, CI 40% to 64%) or for CK-MB activity (57%, CI 43% to 65%) ($P<0.05$).[37]

In the last few years, larger studies have been done that allow for the proper assessment of both the sensitivity and specificity of FABP for AMI diagnosis. In a (single-center) study with 165 patients admitted 3.5 hours (median value) after the onset of chest pain, Ishii et al.[36] found in admission blood samples diagnostic sensitivities and specificities for FABP (>12 μg/L) of 82% and 86%, respectively, and for myoglobin (>105 μg/L) of 73% and 76%, respectively (FABP versus myoglobin significantly different; $P<0.05$). A similar superior performance of

FABP over myoglobin, in terms of both sensitivity and specificity of AMI diagnosis, was also observed in a prospective multicenter study consisting of four European hospitals and including 312 patients admitted 3.3 hours (median value; range 1.5 to 8 hours) after the onset of chest pain suggestive of AMI (EUROCARDI Multicenter Trial).[41,42] For instance, specificities greater than 90% were reached for FABP at 10 μg/L and for myoglobin at 90 μg/L. Using these upper reference concentrations in the subgroup of patients admitted within 3 hours after the onset of symptoms (n = 148), the diagnostic sensitivity of the first blood sample taken was 48% for FABP and 37% for myoglobin, whereas for patients admitted 3 to 6 hours after AMI (n = 86), the sensitivity was 83% for FABP and 74% for myoglobin.[41,42] In addition, the areas under the receiver operating characteristic (ROC) curves, constructed for the admission blood samples from all patients, were 0.901 for FABP and 0.824 for myoglobin (significantly different; $P<0.001$). This better performance of FABP over myoglobin for the early diagnosis of AMI has also been reported in other smaller studies.[24,43–46]

In some of these above-mentioned studies, investigators evaluated whether the diagnostic performance of FABP as an early plasma marker of myocardial injury could further improve when the criterion of a plasma myoglobin:FABP ratio less than 10 (or <14), ie, the exclusion of skeletal muscle as a source of FABP, is taken as an additional parameter.[36,42,46] In each of these study populations, there were a few cases in which both myoglobin and FABP were elevated in the admission plasma sample, but in which the myoglobin:FABP ratio was greater than 10 (or 14). Without this latter result, these patients would be falsely diagnosed as having had myocardial injury. However, because the prevalence of skeletal muscle injury in these study populations was very low (<1% of cases), this additional parameter did not significantly alter the ROC curve for FABP. Therefore, the routine measurement of the myoglobin:FABP ratio in samples from patients suspected for MI does not seem justified. In addition, the myoglobin:FABP ratio cannot provide absolute cardiac specificity.[3]

At first sight it may be surprising that FABP appears as an earlier marker for AMI detection than does myoglobin, even though the two proteins show similar plasma release curves. However, these findings can be explained when realizing that the myocardial content of FABP (0.57 mg/g wet weight) is four- to five-fold lower than that of myoglobin (2.7 mg/g wet weight), yet the plasma reference concentration of FABP (1.8 μg/L) is 19-fold lower than that of myoglobin (34 μg/L) (Table 1). This means that after injury the tissue-to-plasma gradient is almost five-fold steeper for FABP than for myoglobin, making plasma FABP rise above its upper reference concentration at an earlier point after AMI onset than does plasma myoglobin, thereby permitting an earlier diagnosis of AMI.

It is now firmly documented that those patients with unstable angina pectoris who show a significantly increased plasma concentration of TnT (>0.2 μg/L) have a prognosis as serious as do patients with definite AMI.[47,48] This observation most likely relates to the occurrence in these patients of minor myocardial cell necrosis. In those patients in whom unstable angina pectoris is in fact acute minor MI, the advantage of FABP for early assessment of injury may be used. Recently, Katrukha et al.[49] measured FABP and TnI in serial plasma samples from 31 patients with unstable angina and showed that in the admission sample TnI was elevated (cut-off value 0.2 μg/L) in 13% and FABP (cut-off value 6 μg/L) in 54% of patients, whereas at 6 hours after admission TnI was elevated in 58% and FABP in 52% of patients. Importantly, all patients who had an elevated FABP concentration at 6 hours showed an elevated TnI value at 12 hours after admission.[49] These preliminary data suggest that FABP may identify (acute) minor myocardial injury with similar sensitivity as TnI, but at an earlier point after the admission of the patient.

Early Estimation of Myocardial Infarct Size

Myocardial infarct size is commonly estimated from the serial measurement of cardiac proteins in plasma and the calculation of the cumulative release over time (plasma curve area), taking into account the elimination rate of the protein from plasma.[50] This approach requires that the proteins are completely released from the heart after AMI and recovered quantitatively in plasma. Complete recovery is well documented for CK, LDH, and myoglobin (but does not apply for the structural proteins TnT and TnI[33]), and could also be shown for FABP.[31,51] Figure 2 (lower panel) represents the cumulative release patterns of these four proteins, expressed in gram-equivalents (g-eq) of healthy myocardium per liter of plasma (ie, infarct size). The release of FABP and myoglobin is completed much earlier than that of either CK or LDH, but despite this kinetic difference for each of the proteins, the released total quantities yield comparable estimates of the mean extent of myocardial injury when evaluated at 72 hours after the onset of AMI (Fig. 2).

This method to estimate infarct size has proven its value when applied to the evaluation of early thrombolytic therapy in patients with AMI.[52] With the (classically used) enzymatic markers, the method has the drawback that the data on infarct size in the individual patient become available relatively late (72 hours), ie, too late to have influence on acute care.[53] For the more rapidly released markers FABP and myoglobin, infarct size estimation for individual patients is hampered by the fact that these proteins are cleared by the kidneys, and the patients often suffer from renal insufficiency, which would lead to the overestimation of infarct size. De Groot et al.[54] recently suggested the use of individually estimated clearance rates for FABP and myoglobin to measure myocardial infarct size within 24 hours. These individual clearance rates are calculated using glomerular filtration rates (estimated from plasma creatinine concentrations and corrected for age and gender) and plasma volume (corrected for age and gender). This implies that a reliable estimate of myocardial infarct size becomes available when the patient is still in the acute care department if frequent blood samples are taken and analyzed rapidly.

New Approaches to Further Increase the Diagnostic Performance of FABP

A limitation of the use of markers of cell necrosis for assessment of tissue injury is the time lag between the onset of necrosis and the appearance of the marker proteins in plasma. This explains why up to 2 to 3 hours after the onset of AMI, the performance of such markers generally is insufficient for clinical decision making. Therefore, approaches have been presented to further increase the diagnostic performance of the plasma markers in these early hours after AMI.

To circumvent the problem of the upper reference concentration that is defined for populations and used for individual cases, it has been suggested to collect two (or more) serial blood samples during the first hours after admission and to express the difference in marker concentration or activity in these samples. This approach has been applied especially to identify low-risk patients who would show no ECG abnormalities as well as two negative results for protein markers (hence, no significant change with time), and for whom early discharge would be a safe option.[55] In a second EUROCARDI Multicenter Trial, we studied whether in patients admitted for suspected AMI without ECG changes AMI can be ruled out by assay of FABP, myoglobin, or CK-MB mass in two serial blood samples collected upon admission and 1 to 3 hours thereafter, respectively. For comparison, TnT was measured in a third sample taken 12 to 36 hours after admission. Preliminary results from this study revealed that two negative marker concentrations within 3 hours from admission ruled out AMI with very high negative predictive values (>90%) and with the highest value found for FABP (negative predictive value 98%), being similar to that of TnT elevation (≥0.1 μg/L) in the sample taken 12 to 36 hours after admission (B. Haastrup et al., unpublished data, 1999). These data indicate the excellent utility of FABP for early triage and risk stratification of patients with chest pain.

Another approach to further increase the diagnostic performance of FABP in the early hours after the onset of chest pain is to use it in combination with markers of activated blood coagulation.[56] Because intracoronary formation of blood clots on ruptured arteriosclerotic plaques is considered the main cause of AMI, detection of activated blood coagulation potentially allows for the early diagnosis of AMI. Various (small-size) studies have indicated that in the very early hours (0 to 3 hours) after AMI onset, coagulation markers show a higher sensitivity and specificity for AMI detection than do necrosis markers.[57,58] In addition, a tendency toward higher marker concentrations was observed for shorter hospital delays, a finding related to the fact that the acute thrombotic event precedes coronary occlusion and muscle necrosis. In a recent pilot study consisting of 25 patients with either AMI or unstable angina pectoris, we showed that combining a marker of muscle cell necrosis (FABP) and a marker of activated blood coagulation (thrombus precursor protein [TpP]) yielded a markedly higher sensitivity and specificity for AMI detection than either of the markers alone (Fig. 5). Moreover, the performance of such a combined test is expected to be relatively insensitive to hospital delay because TpP will perform better in patients who are admitted earlier, whereas FABP will perform better in patients who are admitted later.[32,58]

Other Applications of the Plasma Marker FABP

FABP was also found to be useful for the early detection of postoperative myocardial tissue loss in patients undergoing coronary bypass surgery.[3,59,60] In these patients, myocardial injury may be caused by global ischemia/reperfusion and, additionally, by postoperative MI. In our study, we found that in such patients plasma CK, myoglobin, and FABP are already significantly elevated 0.5 hours after reperfusion. In the patients who developed postoperative MI, a second increase was observed for each plasma marker protein, but a significant increase was recorded earlier for FABP (4 hours after reperfusion) than for CK or myoglobin (8 hours after reperfusion).[59] These data suggest that FABP would allow for an earlier exclusion of postoperative MI, thus permitting the earlier transfer of these patients from the intensive care unit to the ward. Recently, Hayashida et al.[60] also reported that FABP is an early and sensitive marker for the diagnosis of myocardial injury in patients undergoing cardiac surgery.

The application of FABP as a plasma marker for the early detection

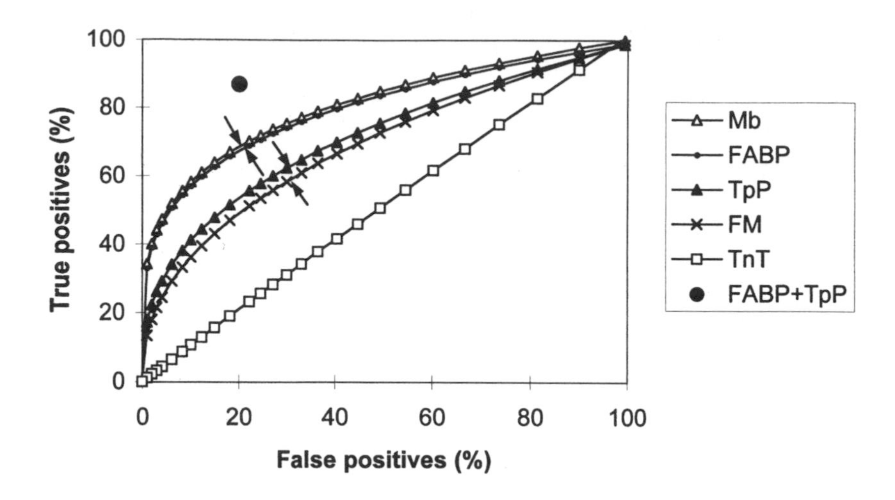

Figure 5. Receiver operating characteristic (ROC) curves for detection of acute myocardial infarction (AMI) in patients having either AMI (n = 15) or unstable angina pectoris (n = 10), comparing selected markers of muscle necrosis (fatty acid binding protein [FABP], myoglobin [Mb], and troponin T [TnT]) and markers of activated blood coagulation (fibrin monomers [FM] and thrombus precursor protein [TpP]). Median hospital delay was 2.8 hours (range 0.8 to 6 hours). ROC curves were obtained from double logarithmic plots. Lack of discrimination by TnT is apparent from its coincidence with the line of identity. Arrows indicate optimal cut-off values. For a combined test, that is when either FABP is greater than 6 μg/L or TpP is greater than 7 mg/L as diagnostic for AMI, the sensitivity was 87% and the specificity 80%. Adapted from Reference 56.

of successful coronary reperfusion in patients with AMI has been investigated by two groups.[61,62] Ishii et al.[61] studied 45 patients treated with intracoronary thrombolysis or direct percutaneous transluminal coronary angioplasty, in whom coronary angiography was performed every 5 minutes to identify the onset of reperfusion. Both plasma FABP and myoglobin were found to rise sharply after the onset of reperfusion, and the relative first-hour increase rates of both markers showed a predictive accuracy of greater than 93%. In a multicenter study consisting of 129 patients with confirmed AMI and receiving thrombolytic agents and in whom the patency of the infarct-related artery was determined from a single-point angiogram, De Groot et al.[62] also observed that FABP and myoglobin perform equally well as markers to discriminate between reperfused and nonreperfused patients. However, these investigators found lower sensitivities and specificities (approximately 70%), which could be improved (to approximately 80%) by normalization to infarct size.[62] These data indicate the equal suitabilities of FABP and myoglobin as noninvasive reperfusion markers, especially in retrospective studies where infarct size is known.

Antibodies directed against FABP have been shown to be useful for the immunohistochemical detection of very recent MIs.[63,64] Partial depletion of FABP was observed in cardiomyocytes with a postinfarction interval of less than 4 hours,[63] indicating that FABP immunostaining can confirm the clinical diagnosis or suspicion of early MI in routine autopsy pathology.

Finally, besides the application of FABP in the early diagnosis of myocardial injury in patients, the marker is now also applied for evaluating MI after coronary artery ligation and for estimating infarct size in experimental animals such as mice and rats.[65–68]

Conclusion

The early diagnosis of acute coronary syndromes is important because it may improve patient treatment and reduce complications. Biochemical markers of myocardial cell damage continue to be important tools for differentiating patients with AMI from those without AMI because specific ST-segment changes in the admission ECG remain absent in a great number of patients with AMI.[1,3] FABP is a novel biochemical marker that shows release characteristics from injured myocardium and elimination rates from plasma that are similar to those of myoglobin, which presently is regarded as the preferred early plasma marker of cardiac injury.[69–71] Experimental studies indicate that this resemblance relates to the similar molecular masses of FABP (14.5 kd) and myoglobin (17.6 kd). Several clinical studies with patients suspected of having AMI reveal a superior performance of FABP over myoglobin (as well as other marker proteins) for the early detection of AMI. This finding most likely relates to marked differences in tissue contents of FABP and myoglobin in cardiac and skeletal muscles that result in a relatively low upper reference concentration in plasma for FABP compared with that for myoglobin. These differences in tissue contents are also reflected in the plasma concentrations of these proteins after either cardiac or skeletal muscle injury, in such a manner that the ratio of the plasma concentrations of myoglobin and FABP can be applied to discriminate myocardial from skeletal muscle injury.

Limitations of the use of FABP as a diagnostic plasma marker in the clinical setting include 1) the relatively small diagnostic window, which extends to only 24 to 30 hours after the onset of chest pain, and 2) its elimination from plasma mainly by renal clearance, possibly causing falsely high values in case of kidney malfunction. These drawbacks can, however, be overcome by the simultaneous measurement in plasma of a late marker such as TnT or TnI and an assay of plasma creatinine to identify patients with renal insufficiency and to calculate a corrected FABP concentration. It is important to note that these same limitations also apply to myoglobin, which is now recommended by both the National Academy of Clinical Biochemistry Committee on Standards of Laboratory Practice[69] and the International Federation of Clinical Chemistry Committee on Standardization of Markers of Cardiac Damage[70] as the preferred early marker of MI, to be used in combination with cardiac TnT or cardiac TnI. In spite of the recognition that, to date, relatively few centers have investigated the performance of FABP for early diagnosis of AMI, the uniformly observed superiority of FABP over myoglobin indicates that the optimal set of biochemical markers of muscle necrosis for assessment of acute coronary syndromes may be FABP together with cardiac TnT or cardiac TnI.[72]

References

1. Adams JE, Abendschein DR, Jaffe AS. Biochemical markers of myocardial injury. Is MB creatine kinase the choice for the 1990s? *Circulation* 1993;88: 750–763.
2. Christenson RH, Azzazy HME. Biochemical markers of the acute coronary syndromes. *Clin Chem* 1998;44:1855–1864.
3. Mair J. Progress in myocardial damage detection: New biochemical markers for clinicians. *Crit Rev Clin Lab Sci* 1997;34:1–66.
4. Glatz JFC, Van der Vusse GJ. Cellular fatty acid-binding proteins. Their function and physiological significance. *Prog Lipid Res* 1996;35:243–282.
5. Banaszak L, Winter N, Xu Z, et al. Lipid binding proteins: A family of fatty acid and retinoid transport proteins. *Adv Protein Chem* 1994;45:89–151.
6. Schaap FG, Specht B, Van der Vusse GJ, et al. One-step purification of rat heart-type fatty acid-binding protein expressed in *Escherichia* coli. *J Chromatogr* 1996;179:61–67.
7. Schreiber A, Specht B, Pelsers MMAL, et al. Recombinant human heart-type fatty acid-binding protein as standard in immunochemical assays. *Clin Chem Lab Med* 1998;36:283–288.
8. Van Breda E, Keizer HA, Vork MM, et al. Modulation of fatty acid-binding protein content of rat heart and skeletal muscle by endurance training and testosterone treatment. *Eur J Physiol* 1992;421:274–279.
9. Glatz JFC, Van Breda E, Keizer HA, et al. Rat heart fatty acid-binding protein content is increased in experimental diabetes. *Biochem Biophys Res Commun* 1994;199:639–646.
10. Vork MM, Trigault N, Snoeckx LHEH, et al. Heterogeneous distribution of fatty acid-binding protein in the hearts of Wistar Kyoto and spontaneously hypertensive rats. *J Mol Cell Cardiol* 1992;24:317–321.
11. Kragten JA, Van Nieuwenhoven FA, Van Dieijen-Visser MP, et al. Distribution of myoglobin and fatty acid-binding protein in human cardiac autopsies. *Clin Chem* 1996;42:337–338.
12. Schaap FG, Binas B, Danneberg H, et al. Impaired long-chain fatty acid

utilization by cardiac myocytes isolated from mice lacking the heart-type fatty acid binding protein gene. *Circ Res* 1999;85:329–337.

13. Glatz JFC, Börchers T, Spener F, et al. Fatty acids in cell signaling: Modulation by lipid binding proteins. *Prostaglandins Leukot Essent Fatty Acids* 1995; 52:121–127.
14. Van der Lee KAJM, Vork MM, De Vries JE, et al. Long-chain fatty acid-induced changes in gene expression in neonatal cardiac myocytes. *J Lipid Res* 2000;41:41–47.
15. Van der Vusse GJ, Glatz JFC, Stam HCG, et al. Fatty acid homeostasis in the normoxic and ischemic heart. *Physiol Rev* 1992;72:881–940.
16. Wodzig KWH, Pelsers MMAL, Van der Vusse GJ, et al. One-step enzyme-linked immunosorbent assay (ELISA) for plasma fatty acid-binding protein. *Ann Clin Biochem* 1997;34:263–268.
17. Tanaka T, Hirota Y, Sohmiya K, et al. Serum and urinary human heart fatty acid-binding protein in acute myocardial infarction. *Clin Biochem* 1991;24: 195–201.
18. Kleine AH, Glatz JFC, Van Nieuwenhoven FA, et al. Release of heart fatty acid-binding protein into plasma after acute myocardial infarction in man. *Mol Cell Biochem* 1992;116:155–162.
19. Ohkaru Y, Asayama K, Ishii H, et al. Development of a sandwich enzyme-linked immunosorbent assay for the determination of human heart type fatty acid-binding protein in plasma and urine by using two different monoclonal antibodies specific for human heart fatty acid-binding protein. *J Immunol Methods* 1995;178:99–111.
20. Knowlton AA, Burrier RE, Brecher P. Rabbit heart fatty acid-binding protein. Isolation, characterization, and application of a monoclonal antibody. *Circ Res* 1989;165:981–988.
21. Katrukha A, Bereznikova A, Filatov V, et al. Development of sandwich time-resolved immunofluorometric assay for the quantitative determination of fatty acid-binding protein (FABP). *Clin Chem* 1997;43:S106. Abstract.
22. Roos W, Eymann E, Symannek M, et al. Monoclonal antibodies to human heart fatty acid-binding protein. *J Immunol Methods* 1995;183:149–153.
23. Robers M, Van der Hulst FF, Fischer MAJG, et al. Development of a rapid microparticle-enhanced turbidimetric immunoassay for plasma fatty acid-binding protein, an early marker of acute myocardial infarction. *Clin Chem* 1998;44:1564–1567.
24. Sanders GT, Schouten Y, De Winter RJ, et al. Evaluation of human heart type fatty acid-binding protein assay for early detection of myocardial infarction. *Clin Chem* 1998;44:A132. Abstract.
25. Siegmann-Thoss C, Renneberg R, Glatz JFC, et al. Enzyme immunosensor for diagnosis of myocardial infarction. *Sensors Actuators* 1996;0:71–76.
26. Schreiber A, Feldbrügge R, Key G, et al. An immunosensor based on disposable electrodes for rapid estimation of fatty acid-binding protein, an early marker of myocardial infarction. *Biosens Bioelectron* 1997;12:1131–1137.
27. Key G, Schreiber A, Feldbrügge R, et al. Multicenter evaluation of an amperometric immunosensor for plasma fatty acid-binding protein: An early marker for acute myocardial infarction. *Clin Biochem* 1999;32:229–231.
28. Robers M, Rensink IJAM, Hack CE, et al. A new principle for rapid immunoassay of proteins based on in situ precipitate-enhanced ellipsometry. *Biophys J* 1999;76:2769–2776.
29. Glatz JFC, Van Bilsen M, Paulussen RJA, et al. Release of fatty acid-binding protein from isolated rat heart subjected to ischemia and reperfusion or to the calcium paradox. *Biochim Biophys Acta* 1988;961:148–152.

30. Tsuji R, Tanaka T, Sohmiya K, et al. Human heart-type cytoplasmic fatty acid-binding protein in serum and urine during hyperacute myocardial infarction. *Int J Cardiol* 1993;41:209–217.
31. Wodzig KWH, Kragten JA, Hermens WT, et al. Estimation of myocardial infarct size from plasma myoglobin or fatty acid-binding protein. Influence of renal function. *Eur J Clin Chem Clin Biochem* 1997;35:191–198.
32. Van Nieuwenhoven FA, Kleine AH, Wodzig KWH, et al. Discrimination between myocardial and skeletal muscle injury by assessment of the plasma ratio of myoglobin over fatty acid-binding protein. *Circulation* 1995;92: 2848–2854.
33. Kragten JA, Hermens WT, Van Dieijen-Visser MP. Cardiac troponin T release into plasma after acute myocardial infarction: Only fractional recovery compared with enzymes. *Ann Clin Biochem* 1996;33:314–223.
34. Hermens WT. Mechanisms of protein release from injured heart muscle. *Dev Cardiovasc Med* 1998;205:85–98.
35. Van Nieuwenhoven FA, Musters RJP, Post JA, et al. Release of proteins from isolated neonatal rat cardiac myocytes subjected to simulated ischemia or metabolic inhibition is independent of molecular mass. *J Mol Cell Cardiol* 1996;28:1429–1434.
36. Ishii J, Wang JH, Naruse H, et al. Serum concentrations of myoglobin vs human heart-type cytoplasmic fatty acid-binding protein in early detection of acute myocardial infarction. *Clin Chem* 1997;43:1372–1378.
37. Glatz JFC, Van der Vusse GJ, Simoons M, et al. Fatty acid-binding protein and the early detection of acute myocardial infarction. *Clin Chim Acta* 1998; 272:87–92.
38. Pelsers MMAL, Chapelle JP, Knapen M, et al. Influence of age and sex and day-to-day and within-day biological variation on plasma concentrations of fatty acid-binding protein and myoglobin in healthy subjects. *Clin Chem* 1999;45:441–443.
39. Yoshimoto K, Tanaka T, Somiya K, et al. Human heart-type cytoplasmic fatty acid-binding protein as an indicator of acute myocardial infarction. *Heart Vessels* 1995;10:304–309.
40. Górski J, Hermens WT, Borawski J, et al. Increased fatty acid-binding protein concentration in plasma of patients with chronic renal failure. *Clin Chem* 1997;43:193–195.
41. Kristensen SR, Haastrup B, Hørder M, et al. Fatty acid-binding protein: A new early marker of AMI. *Scand J Clin Lab Invest* 1996;56(suppl)225:36–37. Abstract.
42. Glatz JFC, Haastrup B, Hermens WT, et al. Fatty acid-binding protein and the early detection of acute myocardial infarction: The EUROCARDI multicenter trial. *Circulation* 1997;96:I215. Abstract.
43. Okamoto F, Tanaka T, Sohmiya K, et al. Heart type fatty acid-binding protein as a new biochemical marker for acute myocardial infarction. *J Mol Cell Cardiol* 1997;29:A306. Abstract.
44. Panteghini M, Bonora R, Pagani F, et al. Heart fatty acid-binding protein in comparison with myoglobin for the early detection of acute myocardial infarction. *Clin Chem* 1997;43:S157. Abstract.
45. Okamoto F, Tanaka T, Sohmiya K, et al. Heart type fatty acid-binding protein as a new biochemical marker for acute myocardial infarction. *J Mol Cell Cardiol* 1997;29:A306. Abstract.
46. Abe S, Saigo M, Yamashita T, et al. Heart fatty acid-binding protein is useful in early and myocardial-specific diagnosis of acute myocardial infarction. *Circulation* 1996;94:I323. Abstract.

47. Hamm CW, Ravkilde J, Gerhardt W, et al. The prognostic value of serum troponin T in unstable angina. *N Engl J Med* 1992;327:146–150.
48. Ravkilde J, Hørder M, Gerhardt W. Diagnostic performance and prognostic value of serum troponin T in suspected acute myocardial infarction. *Scand J Clin Lab Invest* 1993;53:677–683.
49. Katrukha A, Bereznekiva A, Filatov V, et al. Improved detection of minor ischemic cardiac injury in patients with unstable angina by measurement of cTnI and fatty acid binding protein (FABP). *Clin Chem* 1999;45:A139. Abstract.
50. Hermens WT, Van der Veen FH, Willems GM, et al. Complete recovery in plasma of enzymes lost from the heart after permanent coronary occlusion in the dog. *Circulation* 1990;81:649–659.
51. Glatz JFC, Kleine AH, Van Nieuwenhoven FA, et al. Fatty acid-binding protein as a plasma marker for the estimation of myocardial infarct size in humans. *Br Heart J* 1994;71:135–140.
52. Simoons ML, Serruys PW, Van den Brand M, et al. Early thrombolysis in acute myocardial infarction: Limitation of infarct size and improved survival. *J Am Coll Cardiol* 1986;7:717–728.
53. Van der Laarse A. Rapid estimation of myocardial infarct size. *Cardiovasc Res* 1999;44:247–248.
54. De Groot MJM, Wodzig KWH, Simoons ML, et al. Measurement of myocardial infarct size from plasma fatty acid-binding protein or myoglobin, using individually estimated clearance rates. *Cardiovasc Res* 1999;44:315–324.
55. Noble MIM. Can negative results for protein markers of myocardial damage justify discharge of acute chest pain patients after a few hours in hospital? *Eur Heart J* 1999;20:925–927.
56. Hermens WT, Pelsers MMAL, Mullers-Boumans ML, et al. Combined use of markers of muscle necrosis and fibrinogen conversion in the early differentiation of myocardial infarction and unstable angina. *Clin Chem* 1998;44: 890–892.
57. Merlini PA, Bauer KA, Oltrona L, et al. Persistent activation of coagulation mechanism in unstable angina and myocardial infarction. *Circulation* 1994; 90:61–68.
58. Carville DGM, Dimitrijevic N, Walsh M, et al. Thrombus precursor protein (TpP): Marker of thrombosis early in the pathogenesis of myocardial infarction. *Clin Chem* 1996;42:1537–1541.
59. Fransen EJ, Maessen JG, Hermens WT, et al. Demonstration of ischemia-reperfusion injury separate from postoperative infarction in coronary artery bypass graft patients. *Ann Thorac Surg* 1998;65:48–53.
60. Hayashida N, Chihara S, Akasu K, et al. Plasma and urinary levels of heart fatty acid-binding protein in patients undergoing cardiac surgery. *Jpn Circ J* 2000;64:18–22.
61. Ishii J, Nagamura Y, Nomura M, et al. Early detection of successful coronary reperfusion based on serum concentration of human heart-type cytoplasmic fatty acid-binding protein. *Clin Chim Acta* 1997;262:13–27.
62. De Groot MJM, Simoons ML, Hermens WT, et al. Assessment of coronary reperfusion in patients with myocardial infarction using fatty acid-binding protein concentrations in plasma. *Clin Chem Lab Med*, in press Abstract.
63. Kleine AH, Glatz JFC, Havenith MG, et al. Immunohistochemical detection of very recent myocardial infarctions in man with antibodies against heart type fatty acid-binding protein. *Cardiovasc Pathol* 1993;2:63–69.
64. Watanabe K, Wakabayashi H, Veerkamp JH, et al. Immunohistochemical

distribution of heart-type fatty acid-binding protein immunoreactivity in normal human tissues and in acute myocardial infarct. *J Pathol* 1993;170: 59–65.

65. Knowlton AA, Apstein CS, Saouf R, et al. Leakage of heart fatty acid binding protein with ischemia and reperfusion in the rat. *J Mol Cell Cardiol* 1989; 21:577–583.
66. Volders PGA, Vork MM, Glatz JFC, et al. Fatty acid-binding proteinuria diagnosis myocardial infarction in the rat. *Mol Cell Biochem* 1993;123: 185–190.
67. Sohmiya K, Tanaka T, Tsuji R, et al. Plasma and urinary heart-type cytoplasmic fatty acid-binding protein in coronary occlusion and reperfusion induced myocardial injury model. *J Mol Cell Cardiol* 1993;25:1413–1426.
68. Aartsen WM, Pelsers MMAL, Hermens WT, et al. Heart fatty acid binding protein and cardiac troponin T plasma concentrations as markers for myocardial infarction after coronary artery ligation in mice. *Eur J Physiol* 2000; 439:416–422.
69. Wu AHB, Apple FA, Gibler WB, et al. National Academy of Clinical Biochemistry Standards on Laboratory Practice: Recommendations for the use of cardiac markers in coronary artery diseases. *Clin Chem* 1999;45: 1104–1121.
70. Panteghini M, Apple FS, Christenson RH, et al. Use of biochemical markers in acute coronary syndromes. IFCC Scientific Division, Committee on Standardization of Markers of Cardiac Damage. *Clin Chem Lab Med* 1999; 37:687–693.
71. Storrow AB, Gibler WB. The role of cardiac markers in the emergency department. *Clin Chim Acta* 1999;284:187–196.
72. Wu AH. Analytical and clinical evaluation of new diagnostic tests for myocardial damage. *Clin Chim Acta* 1998;272:11–21.
73. Young AC, Scapin G, Kromminga A, et al. Structural studies on human muscle fatty acid binding protein at 1.4 A resolution: Binding interactions with three C18 fatty acids. *Structure* 1994;2:523–534.
74. Sacchettini JC, Gordon JI, Banaszak LJ. Refined apoprotein structure of rat intestinal fatty acid binding protein produced in Escherichia coli. *Proc Natl Acad Sci U S A* 1989;86:7736–7740.
75. Van Nieuwenhoven FA. *Heart Fatty Acid-Binding Proteins. Role in Cardiac Fatty Acid Uptake and Marker for Cellular Damage.* Thesis, Maastricht University; 1996:65–71.

Chapter 14

Oxidized Low-Density Lipoprotein and Malondialdehyde-Modified Low-Density Lipoprotein: *Markers of Coronary Artery Disease*

Paul Holvoet, PhD, Frans Van de Werf, MD, PhD, Johan Vanhaecke, MD, PhD, and Désiré Collen, MD, PhD

Introduction

Lipid oxidation of low-density lipoprotein (LDL) may be initiated by metal ions, possibly in association with phospholipase activity, or may be catalyzed by myeloperoxidase, independent of metal ions. Lipid oxidation results in the generation of aldehydes that substitute lysine residues in the apolipoprotein B-100 moiety and thus in the oxidative modification of LDL. The resulting oxidatively modified LDL is generally referred to as oxidized LDL. Both lipid and protein oxidation of LDL can occur.

Endothelial injury, associated with the increased production of free radicals during oxidative stress, is associated with increased prostaglandin synthesis and platelet adhesion/activation. These processes are associated with the release of aldehydes that induce oxidative modification of the protein moiety of LDL in the absence of lipid oxidation and

Supported in part by a grant from the Fonds voor Geneeskundig Wetenschappelijk Onderzoek-Vlaanderen (Project 7.0022.98) and by the Interuniversitaire Attractiepolen (Program 4/34). J. Vanhaecke holds the Michael Ondetti Chair in Cardiology at the University of Leuven School of Medicine.

From: Adams JE III, Apple FS, Jaffe AS, Wu AHB (eds). *Markers in Cardiology: Current and Future Clinical Applications.* Armonk, NY: Futura Publishing Company, Inc.; © 2001.

thus in the generation of what is referred to as malondialdehyde (MDA)-modified LDL.

Although the association between oxidative modification of LDL and cardiovascular disease has been suspected for a long time, there was no definitive proof primarily because of the lack of assays that allowed for the direct quantification of oxidatively modified LDL in the blood. Recently, we have developed sensitive and specific assays for oxidized LDL and MDA-modified LDL.

We have demonstrated an association between coronary artery disease (CAD) and increased plasma levels of oxidized LDL. The increase of circulating oxidized LDL is most probably independent of plaque instability. Indeed, plasma levels of oxidized LDL were very similar for patients with stable CAD and for patients with acute coronary syndromes. Recently, a prospective study in cardiac transplant patients suggested an active role of oxidized LDL in the development of CAD. Oxidized LDL may contribute to the progression of atherosclerosis by enhancing endothelial injury and by inducing foam cell generation and smooth muscle proliferation.

We also demonstrated that acute coronary syndromes were associated with an increased release of MDA-modified LDL that was independent of the necrosis of myocardial cells. These data suggest that oxidized LDL is a marker of coronary arteriosclerosis, whereas MDA-modified LDL is a marker of plaque instability.

Oxidative Modification of LDL

Table 1 summarizes different possible mechanisms of oxidative modification of LDL. Metal-ion induced, in vitro oxidation of LDL appears to occur in three phases. During the initial lag-phase, endogenous LDL antioxidants (eg, vitamin E) are consumed. During the propagation phase, unsaturated fatty acids are rapidly oxidized to lipid hydroperoxides. During the decomposition phase, hydroperoxides are converted to reactive aldehydes (eg, MDA and 4-hydroxynonenal).[1] Endothelial cells, monocytes, macrophages, lymphocytes, and smooth muscle cells are all capable of enhancing the rate of oxidation of LDL.[2] During inflammation, several cell types synthesize and secrete phospholipase A_2. Myeloperoxidase, a heme protein secreted by activated phagocytes, oxidizes L-tyrosine to a tyrosyl radical that is a physiological catalyst for the initiation of lipid oxidation in LDL. In striking contrast to other cell-mediated mechanisms for LDL oxidation, the myeloperoxidase-catalyzed reaction is independent of free metal ions.[3] Lipid oxidation results in the generation of aldehydes that substitute lysine residues in the apolipoprotein B-100 moiety of LDL and causes its B-

Table 1

Overview of Possible Mechanisms of Oxidative Modification of LDL

Oxidative stress in Endothelial cells Macrophages	Activation of Macrophages	Prostaglandin synthesis by endothelial cells Platelet activation
↓	↓	↓
Radicals Lipoxygenase/ Phospholipase	Myeloperoxidase Activity	
↓	↓	
Consumption of Antioxidants, lipid Peroxidation in LDL	Protein and lipid Oxidation in LDL	Aldehyde production Independent of lipid Oxidation in LDL
↓	↓	↓
Aldehyde substitution of Lysine residues in APO B-100	Aldehyde substitution of lysine residues in APO B-100	Aldehyde substitution of Lysine residues in APO B-100
↓	↓	↓
Oxidized LDL	Oxidized LDL	MDA-modified LDL

APO = apolipoprotein; LDL = low-density lipoprotein; MDA = malondialdehyde.

100 fragmentation. The resulting oxidatively modified LDL is generally referred to as oxidized LDL. An mAb-4E6-based competition enzyme-linked immunosorbent assay (ELISA) may be used for the measurement of oxidized LDL in plasma.[4] The C_{50} values, concentrations that are required to obtain 50% inhibition of antibody binding in the ELISA, are 25 mg/dL for native LDL and 0.25 mg/dL for oxidized LDL with at least 60 aldehyde-substituted lysines per apoB-100.

Oxidative stress in endothelial cells and platelet activation are associated with the oxidation of arachidonic acid to aldehydes. These interact with lysine residues in the apolipoprotein B-100 moiety of LDL, resulting in oxidative modification of the protein moiety of LDL in the absence of lipid oxidation.[5–7] The resulting oxidatively modified LDL is generally referred to as MDA-modified LDL.[8] An mAb-1H11-based competition ELISA may be used for the measurement of MDA-modified LDL in plasma. The C_{50} values are 25 mg/dL for native LDL and for oxidized LDL and 0.25 mg/dL for MDA-modified LDL with at least 60 aldehyde-substituted lysines per apoB-100.

Oxidatively Modified LDL as a Marker of Atherosclerotic Cardiovascular Disease

The association of oxidative modification of LDL with stable angina and acute coronary syndromes has been studied. Table 2 shows

Table 2

Characteristics of Controls, CAD Patients, and Heart Transplant Patients

	Controls (n = 65)	Stable Angina (n = 64)	Unstable Angina (n = 42)	AMI (n = 62)	Transplant Patients without CAD (n = 79)	Transplant Patients with CAD (n = 28)
Age	52 ± 10	65 ± 10^b	72 ± 12^a	63 ± 12	56 ± 11	58 ± 8.4
Male/female ratio	31/34	53/11	28/14	45/17	$73/6^c$	$22/6^a$
Total cholesterol (mg/dL)	180 ± 37	177 ± 35	174 ± 37	178 ± 44	193 ± 37	182 ± 36
LDL cholesterol (mg/L)	105 ± 43	115 ± 30	109 ± 33	112 ± 40	117 ± 34	103 ± 29
HDL cholesterol (mg/dL)	48 ± 22	38 ± 13	45 ± 16	39 ± 12	50 ± 21	53 ± 19
Triglycerides (mg/dL)	130 ± 60	123 ± 46	100 ± 55	128 ± 67	136 ± 71	124 ± 48
Oxidized LDL (mg/dL)	0.71 ± 0.26	2.65 ± 1.36^c	2.84 ± 0.91^c	3.44 ± 1.5^c	1.27 ± 0.61^b	2.49 ± 1.44^c
MDA-modified LDL (mg/dL)	0.37 ± 0.14	0.46 ± 0.20	1.33 ± 0.49^c	1.14 ± 0.46^c	0.38 ± 0.16	0.39 ± 0.23
Troponin I (μg/L)	0.025 ± 0.025	0.035 ± 0.075	0.19 ± 0.44^c	0.68 ± 1.2^c	0.026 ± 0.029	0.041 ± 0.080
C-reactive protein (mg/dL)	3.4 ± 2.6	5.30 ± 6.89	12.48 ± 23.50^b	22 ± 35^c	4.0 ± 5.6	4.1 ± 2.6
D-dimer (μg/dL)	13 ± 6.9	30 ± 28	37 ± 44^a	57 ± 75^c	21 ± 23	21 ± 17

Quantitative data represent mean ± SD. *P* values determined by nonparametric multiple comparison test, except for male/female ratios that were compared by Chi-square analysis. $^aP < 0.05$; $^bP < 0.01$; $^cP < 0.001$. AMI = acute myocardial infarction; CAD = coronary artery disease.

the characteristics of healthy controls and patients with cardiovascular disease. Compared with controls, plasma levels of oxidized LDL were 3.7-fold higher ($P<0.001$) for patients with stable angina pectoris, 4.0-fold higher ($P<0.001$) for patients with unstable angina pectoris, and 4.8-fold higher ($P<0.001$) for patients with acute myocardial infarction (AMI). Plasma levels of MDA-modified LDL, similar for controls and patients with stable angina pectoris, were 3.6-fold higher ($P<0.001$) for patients with unstable angina pectoris and 3.1-fold higher ($P<0.001$) for AMI patients.[9] C-reactive protein levels, similar for controls and patients with stable CAD, were 3.7-fold higher for unstable angina patients and 6.5-fold higher for AMI patients ($P<0.001$). Plasma levels of C-reactive protein were higher ($P<0.05$) for AMI patients than for patients with unstable angina.[9] Troponin I levels, similar for controls and patients with stable CAD, were 5.4-fold higher for patients with unstable angina ($P<0.001$) and 19-fold higher for AMI patients ($P<0.001$) than for controls. AMI patients had 3.6-fold higher plasma troponin I level than patients with unstable angina ($P<0.001$).[9] D-dimer levels were similar for controls and for patients with chronic stable angina or unstable angina pectoris and were 4.4-fold higher for AMI patients.[9]

Multiple regression analysis was performed to evaluate the association between angiographically detected CAD and plasma levels of oxidized LDL and of MDA-modified LDL. After correction for age, gender, LDL cholesterol, and high-density lipoprotein (HDL) cholesterol, CAD was associated with elevated plasma levels of oxidized LDL. Multiple regression analysis was performed on the subgroups of CAD patients to study the association between acute coronary syndromes and plasma levels of oxidized LDL and of MDA-modified LDL. Acute coronary syndromes were associated with increased MDA-modified LDL, troponin I, and C-reactive protein. Differences between plasma levels of oxidized LDL in patients with stable CAD and patients with acute coronary syndromes were less pronounced.[9]

Univariate logistic regression analysis revealed an association of clinically diagnosed acute coronary syndromes with C-reactive protein ($\chi^2=21$; $P<0.0001$), troponin I ($\chi^2=25$; $P<0.0001$), and MDA-modified LDL ($\chi^2=19$; $P=0.0003$). C-reactive protein correlated with MDA-modified LDL ($r=0.32$; $P=0.013$) and troponin I ($r=0.59$; $P=0.0001$); troponin I correlated with C-reactive protein and MDA-modified LDL ($r=0.44$; $P=0.0001$). MDA-modified LDL correlated both with troponin I and with C-reactive protein.

Receiver operating characteristic curve analysis revealed that MDA-modified LDL ($\chi^2=10.2$; $P=0.0014$), but not troponin I and C-reactive protein, could discriminate between stable CAD and unstable angina. In contrast, troponin I ($\chi^2=14.5$; $P=0.0007$), but neither MDA-modified LDL nor C-reactive protein, discriminated between unstable angina and AMI. Both MDA-modified LDL ($\chi^2 = 14.5$; $P = 0.0001$) and troponin I ($\chi^2 = 5.3$; $P = 0.021$), but not C-reactive protein, discriminated between stable CAD and AMI.

Table 3 shows the ratio of positive versus negative tests for C-reactive protein, troponin I, and MDA-modified LDL, respectively, in patients with stable CAD, unstable angina, AMI, and acute coronary syndromes (unstable angina pectoris or AMI), respectively. At a cut-off value of 10 mg/dL (value exceeding the 95th percentile of distribution for patients with stable angina), the sensitivity of C-reactive protein was 19% for unstable angina and 42% for AMI, whereas the specificity was 95%. At a cut-off value of 0.05 μg/L (value exceeding the 95th percentile of distribution for patients with stable angina), the sensitivity of troponin I was 38% for unstable angina and 90% for AMI, whereas the specificity was 95%. At a cut-off value of 0.70 mg/dL (value exceeding the 95th percentile of distribution for patients with stable angina), the sensitivity of MDA-modified LDL was 95% for unstable angina and 95% for AMI, whereas the specificity was 95%.[10]

The study cohort consisted of 98 patients (35 with stable CAD, 18 with unstable angina, and 45 with AMI), who have been described elsewhere,[5] and 70 new patients (29 with stable CAD, 24 with unstable angina, and 17 with AMI). The age, gender, serum levels of total cholesterol, LDL cholesterol, HDL cholesterol, and triglycerides, and the levels of C-reactive protein, troponin I, and MDA-modified LDL of patients with stable angina and acute coronary syndromes, respectively, in the two subgroups were very similar. The sensitivity of troponin I for unstable angina was 44% for the first subgroup and 33% for the second subgroup. The corresponding values for AMI were 93% and 83%, respectively. The sensitivity of MDA-modified LDL for unstable angina was 95% for both subgroups. The corresponding values for AMI were 95% and 83%, respectively.[10]

Recently, the relation between circulating oxidized LDL and major cardiovascular risk factors of age, hypertension, hypercholesterolemia, dyslipidemia, obesity, diabetes, and smoking was assessed. Levels of oxidized LDL were determined in plasma samples obtained from 352 patients admitted to the departments of cardiology and endocrinology of the University Hospital in Leuven: 106 patients with CAD and 246 patients without clinical evidence of CAD. The characteristics of the patients, not included in the previous study, are illustrated in Table 4.

In a multivariate model that did not include circulating oxidized LDL, age ($P < 0.001$), hypercholesterolemia ($P < 0.001$), dyslipidemia ($P = 0.0034$), male sex ($P = 0.0015$), and hypertension ($P = 0.019$) predicted CAD. In a model containing circulating oxidized LDL, plasma levels of oxidized LDL ($P < 0.001$), age ($P < 0.001$), and dyslipidemia

Table 3

Chi-Square Analysis of Ratio of Positive versus Negative Tests for C-Reactive Protein, Troponin I, and MDA-Modified LDL in Patients with Stable CAD and Patients with Acute Coronary Syndromes Clinically Presenting as Unstable Angina or AMI

Parameter	Stable CAD	Unstable Angina			AMI			Acute Coronary Syndromes		
	Ratio	Ratio	χ^2	P	Ratio	χ^2	P	Ratio	χ^2	P
C-Reactive Protein	3/61	8/34	4.2	0.041	26/36	23	<0.0001	34/70	17	<0.0001
Troponin I	3/61	16/26	17	<0.0001	56/6	89	<0.0001	72/32	64	<0.0001
MDA-Modified LDL	3/61	40/2	83	<0.0001	59/3	100	<0.0001	99/5	132	<0.0001

Cut-off values were 10 mg/dL for C-reactive protein, 0.05 μg/L for troponin I, and 0.7 mg/dL for MDA-modified LDL. All these values exceeded the 95th percentile of distribution in individuals with stable CAD. The ratios between positive and negative samples are represented; χ^2 values determined by Yates' continuity corrected Chi-square test, and P values were obtained by comparison with stable CAD patients. AMI = acute myocardial infarction; CAD = coronary artery disease; LDL = low-density lipoprotein; MDA = malondialdehyde.

Table 4

Patient Characteristics

Characteristic	non CAD (n=246)	CAD (n=106)	*P*
Age, y (mean± SD)	44±16	65±9.1	<0.001
Gender (male/female)	92/154	68/38	<0.001
Hypertension	52 (21)	41 (39)	<0.001
Diabetes type 2	47 (19)	30 (28)	0.068
Hypercholesterolemia	46 (19)	53 (50)	<0.001
Treated with statins	6 (2.4)	20 (19)	<0.001
Treated with fibrates	8 (3.3)	7 (6.6)	0.16
Dyslipidemia	37 (15)	33 (22)	<0.001
Smokers	63 (26)	49 (46)	<0.001
BMI (height/weight^2)	28±8.5	27±5.1	0.26
Total cholesterol (mg/dL)	168±34	185±40	<0.001
LDL cholesterol (mg/dL)	96±31	112±32	<0.001
Triglycerides (mg/dL)	121±89	156±91	<0.001
HDL cholesterol (mg/dL)	47±14	42±15	0.0028
Tot-C/HDL-C	3.84±1.66	4.71±1.66	<0.001
Oxidized LDL (mg/dL)	1.11±0.82	2.91±1.15	<0.001

Percentages in parentheses. BMI = body mass index; CAD = coronary artery disease; HDL = high-density lipoprotein; LDL = low-density lipoprotein; tot-/HDL-C = total cholesterol to HDL cholesterol ratio.

($P=0.0024$) predicted CAD (Table 5). Overall, 91% of patients were predicted correctly.

Predictors of circulating oxidized LDL were, besides CAD, hypercholesterolemia ($P<0.001$), obesity ($P<0.001$), male sex ($P=0.006$), dyslipidemia ($P=0.012$), age ($P=0.018$), and hypertension ($P=0.029$) (Tables 6 and 7).

Oxidized LDL and the Development of Transplant CAD

In a cohort study, the association between transplant-associated CAD and the oxidative modification of LDL was studied in heart transplant patients. The characteristics of these patients are summarized in Table 8. Levels of oxidized LDL in heart transplant patients with normal coronary angiograms were 1.8-fold higher than in controls.[11] Levels of MDA-modified LDL, troponin I, C-reactive protein, and D-dimer for transplant patients with normal angiograms were similar to those for controls. Heart transplant patients with angiographically documented

Table 5

Logistic Regression Analysis of Relation of Potential Risk Factors with CAD

Factor	Wald	*P*	OR
Multivariate model 1			
Age	58	<0.001	1.12
Hypercholesterolemia	9.3	<0.001	3.68
Male gender	10	0.0015	2.94
Dyslipidemia	8.6	0.0034	3.04
Hypertension	5.5	0.019	2.25
Constant	27	<0.001	
Multivariate model 2			
Oxidized LDL	47	<0.001	4.92
Age	30	<0.001	1.12
Dyslipidemia	9.3	0.0024	4.15
Hypercholesterolemia	3.6	0.057	2.19
Constant	25	<0.001	

The multivariate model 1 contained age, gender, hypertension, diabetes, hypercholesterolemia, dyslipidemia, smoking, and body mass index as covariates. The R^2 value of this model was 0.57, its sensitivity 74%, and its specificity 88%. Overall, 84% of patients were predicted correctly. The multivariate model 2 contained levels of circulating oxidized LDL and all the other covariates included in the first model. The R^2 value of this model was 0.74%, its sensitivity 82%, its specificity 95%. Overall, 91% of patients were predicted correctly. CAD = coronary artery disease; LDL = low-density lipoprotein; OR = odds ratio; Wald = Wald's statistic.

CAD had two-fold higher plasma levels of oxidized LDL than those heart transplant patients with normal angiograms. Levels of oxidized LDL were similar for patients with transplant CAD and for patients with ischemic heart disease. Transplant CAD patients had levels of MDA-modified LDL, troponin I, C-reactive protein, and D-dimer similar to those of patients with chronic stable angina. Multivariate logistic regression analysis revealed three parameters that significantly and independently correlated with transplant-associated CAD: circulating oxidized LDL, time after transplantation, and age of donor.[11]

The 99 heart transplant patients with normal baseline coronary angiograms were included in a prospective study. During a 2-year follow-up, 21 patients developed transplant CAD (cases), and 78 did not (controls). Baseline plasma levels of oxidized LDL were 2.0–fold higher in cases than in controls ($P<0.001$). Baseline levels of MDA-modified LDL were similar in cases and in controls. At the end of a 2-year follow-up, plasma levels of oxidized LDL were 2.8-fold higher ($P<0.001$) in cases than in controls. During the follow-up, plasma levels of oxidized

Table 6

Predictors of Circulating Oxidized LDL for Total Study Population

Factor	Univariate		Multivariate	
	R	*P*	F Change	*P*
CAD	0.67	<0.001	286	<0.001
BMI	0.12	0.015	16	<0.001
Hypercholesterolemia	0.34	<0.001	9.3	0.002
Male gender	0.24	<0.001	7.5	0.006
Hypertension	0.22	<0.001	4.8	0.029
Age	0.47	<0.001		
Smoking	0.25	<0.001		
Dyslipidemia	0.17	0.001		
Diabetes type 2	0.11	0.021		

R indicates Pearson correlation coefficient. The relation of angiographically assessed CAD, age, gender, hypertension, diabetes, hypercholesterolemia, dyslipidemia, smoking, and BMI with circulating oxidized LDL was assessed by stepwise multivariate regression analysis. The R^2 value of the model was 0.50. CAD = coronary artery disease; BMI = body mass index; LDL = low-density lipoprotein.

Table 7

Predictors of Circulating Oxidized LDL for Patients without Clinical Evidence of CAD

Factor	Univariate		Multivariate	
	R	*P*	F Change	*P*
Hypercholesterolemia	0.28	<0.001	25	<0.001
BMI	0.31	<0.001	18	<0.001
Dyslipidemia	0.27	<0.001	6.5	0.012
Age	0.23	<0.001	5.6	0.018
Diabetes type 2	0.21	<0.001		
Smoking	0.15	0.009		
Male gender	0.10	0.054		

R indicates Pearson correlation coefficient. The relation of age, gender, hypertension, diabetes, hypercholesterolemia, dyslipidemia, smoking, and BMI with circulating oxidized LDL was assessed by stepwise multivariate regression analysis. The R^2 value of the multivariate model was 0.18. CAD = coronary artery disease; BMI = body mass index; LDL = low-density lipoprotein.

Table 8

Relation of Baseline Levels of Oxidized LDL and Transplant CAD with Possible Covariates

	Oxidized LDL		CAD	
Covariate	R^2	*P*	χ^2	*P*
Age	0.0057	0.46	4.8	0.029
Gender	0.0095	0.78	0.03	0.86
Pretransplant history	0.055	0.017	2.49	0.11
Time after transplantation	0.0025	0.68	4.60	0.033
Hypertension	0.00019	0.89	0.12	0.73
Diabetes	0.0013	0.72	0.54	0.46
Peripheral vascular disease	0.0020	0.86	0.08	0.78
Cytomegalovirus infection	0.0012	0.73	0.48	0.49
Rejections	0.0015	0.71	0.050	0.94
Statins	0.011	0.31	3.70	0.054
Fibrates	0.0019	0.66	0.38	0.54
Calcium channel blockers	0.014	0.24	1.44	0.23
Total cholesterol	0.010	0.66	0.16	0.69
LDL cholesterol	0.059	0.014	1.83	0.18
HDL cholesterol	−0.014	0.24	−5.8	0.016
Triglycerides	0.011	0.88	1.05	0.30
Baseline oxidized LDL	—	—	16	<0.0001

The relation of baseline levels of oxidized LDL with covariates was assessed by linear regression. The relation of transplant CAD with covariates was assessed by logistic regression. CAD = coronary artery disease; HDL = high-density lipoprotein; LDL = low-density lipoprotein.

LDL increased 5.4-fold more in cases than in controls ($P<0.001$). Parameters that predicted baseline levels of oxidized LDL or transplant CAD are illustrated in Table 8. Multivariate logistic regression analysis showed that baseline plasma levels of oxidized LDL predicted the development of transplant CAD independent of pretransplant ischemic heart disease or dilated cardiomyopathy, time after transplantation, age, and serum levels of LDL and HDL cholesterol.[11]

Conclusion

Our cohort studies showed that plasma levels of oxidized LDL are significantly elevated in CAD patients. These levels are very similar for patients with stable CAD and for patients with acute coronary syndromes, suggesting that their increase is independent of plaque instability. In contrast, plasma levels of MDA-modified LDL were significantly higher for patients with acute coronary syndromes than for patients with stable CAD, suggesting that increases of plasma levels of MDA-modified LDL are dependent on the ischemic syndromes for patients with unstable angina pectoris or AMI. The association between MDA-modified LDL and troponin I, a marker of ischemic syndromes, further supports this hypothesis.

Our prospective study of heart transplant patients showed that baseline levels of oxidized LDL predicted the development of transplant CAD independent of levels of LDL and HDL cholesterol and of the pretransplant history of ischemic heart disease. Thus, the level of oxidized LDL in the blood was an independent predictor of the development of transplant CAD that was associated with a further increase of plasma levels of oxidized LDL. Although the study identified oxidized LDL as a prognostic marker of transplant CAD, it did not prove that oxidized LDL has an active role in the development of CAD. Oxidized LDL may, however, contribute to the progression of atherosclerosis by enhancing endothelial injury and by inducing foam cell generation and smooth muscle proliferation.

Intervention trials are needed to evaluate the active role of oxidized LDL in the development of CAD in general. The recent finding that plasma levels of oxidized LDL are increased in obese patients and in type II diabetes patients who are at an increased risk for CAD suggests that this may be the case. The elevation of oxidized LDL in these patients could only partly be explained by the characteristic dyslipidemia. Interventions may aim at a further decrease of LDL cholesterol, at an increase of levels of antioxidants in LDL, and/or at an increase of HDL that contains enzymes such as paraoxonase and platelet-activating factor acetylhydrolase that may prevent the oxidation of LDL or may degrade oxidized phospholipids in LDL.

Robust assay kits have yet to be developed to assess the prognostic value of oxidized LDL for the progression of coronary atherosclerosis and the predictive value of MDA-modified LDL for acute coronary syndromes. Therefore, oxidatively modified LDL standards must be stabilized and, preferably, sandwich-type assays must be developed. Preferably, the extent of coronary atherosclerosis should also be assessed by more sensitive methods than angiography, for example by intravascular ultrasound. Finally, the association between oxidatively modified LDL and C-reactive protein measured in a recently established high-sensitivity assay must be assessed.

References

1. Esterbauer H, Jurgens G, Quehenberger Q, Koller E. Autooxidation of human low density lipoprotein: Loss of polyunsaturated fatty acids and vitamin E and generation of aldehydes. *J Lipid Res* 1987;28:495–509.

2. Steinbrecher UP, Parthasarathy S, Leake DS, et al. Modification of low density lipoprotein by endothelial cells involves lipid peroxidation and degradation of low density lipoprotein phospholipids. *Proc Natl Acad Sci U S A* 1984;81:3883–3887.
3. Savenkova ML, Mueller DM, Heinecke JW. Tyrosyl radical generated by myeloperoxidase is a physiological catalyst for the initiation of lipid peroxidation in low density lipoprotein. *J Biol Chem* 1994;269:20394–20400.
4. Holvoet P, Stassen JM, Van Cleemput J, et al. Correlation between oxidized low density lipoproteins and coronary artery disease in heart transplant patients. *Arterioscler Thromb Vasc Biol* 1998;18:100–107.
5. Farber HW, Barnett HF. Differences in prostaglandin metabolism in cultured aortic and pulmonary arterial endothelial cells exposed to acute and chronic hypoxia. *Circ Res* 1991;68:1446–1457.
6. Lynch SM, Morrow JD, Roberts LJ II, Frei B. Formation of non-cyclooxygenase-derived prostanoids (F_2 isoprostanes) in plasma and low density lipoprotein exposed to oxidative stress in vitro. *J Clin Invest* 1994;93:998–1004.
7. Laskey RE, Mathews WR. Nitric oxide inhibits peroxynitrite-induced production of hydroxyeicosatetraenoic acids and F_2-isoprostanes in phosphatidylcholine liposomes. *Arch Biochem Biophys* 1996;330:193–198.
8. Holvoet P, Perez G, Zhao Z, et al. Malondialdehyde-modified low density lipoproteins in patients with atherosclerotic disease. *J Clin Invest* 1995;95: 2611–2619.
9. Holvoet P, Vanhaecke J, Janssens S, et al. Oxidized LDL and malondialdehyde-modified LDL in patients with acute coronary syndromes and stable coronary artery disease. *Circulation* 1998;98:1487–1494.
10. Holvoet P, Collen D, Van de Werf F. Malondialdehyde-modified LDL as a marker of acute coronary syndromes. *JAMA* 1999;281:1718–1721.
11. Holvoet P, Van Cleemput J, Collen D, Vanhaecke J. Oxidized low-density lipoprotein is a prognostic marker of transplant-associated coronary artery disease. *Arterioscler Thromb Vasc Biol* 2000;20:698–702.

Chapter 15

High-Sensitivity C-Reactive Protein:

A Novel Inflammatory Marker for Predicting Risk of Coronary Artery Disease

Paul M. Ridker, MD, MPH

Half of all myocardial infarctions (MIs) occur among individuals without overt hyperlipidemia. Thus, for aggressive primary prevention programs to be successful, novel markers of cardiovascular risk must be found that can 1) easily and inexpensively be measured in outpatient clinical settings; 2) consistently predict future vascular risk in different patient groups; and 3) add to our ability to predict risk over and above standard lipid screening.[1] Accumulating clinical data suggest that measurement of high-sensitivity C-reactive protein (hs-CRP), a marker for low-grade systemic inflammation, may provide such a tool for both primary and secondary screening of patients with cardiovascular disease.[2]

To understand the potential role of hs-CRP in cardiovascular risk prediction, it is first critical to recognize the role of inflammation in the atherothrombotic process. Accumulating evidence from both basic and applied laboratories indicates that inflammation plays a critical role in several stages of atherogenesis, including acute plaque rupture as well as early initiation of foam cell deposits.[3] Indeed, the processes of leukocyte recruitment and attachment to the endothelial wall in early atherogenesis and the accumulation of cytokine and metalloproteinase that produces mononuclear cells at sites of acute plaque rupture both involve a full cascade of inflammatory cells and circulating inflammatory mediators.[4]

From: Adams JE III, Apple FS, Jaffe AS, Wu AHB (eds). *Markers in Cardiology: Current and Future Clinical Applications.* Armonk, NY: Futura Publishing Company, Inc.; © 2001.

The identification of inflammation as a critical component of atherothrombosis has led to the development of several clinical assays capable of detecting this process in both acutely ill and apparently healthy individuals. Markers of the inflammatory process that appear to predict coronary events include hs-CRP, serum amyloid A, fibrinogen, soluble intercellular adhesion molecule type 1 (sICAM-1), and interleukin 6 (IL-6). Of these inflammatory markers, hs-CRP may prove to be the most clinically useful because it is easily measured with commercially available reagents[5] and appears to have the strongest effect as an independent predictor of future vascular events.[6]

From a historical perspective, C-reactive protein (CRP) was first described in 1930 as a circulating inflammatory protein capable of reacting with the somatic C-polysaccharide of *Streptococcus pneumoniae*.[7] Subsequent studies have demonstrated that CRP, a 1.14 kd peptide composed of five identical subunits, is produced almost exclusively by hepatocytes in response to circulating levels of IL-6.[8] Because levels rise several hundred fold in response to acute infections or trauma, the typical clinical use of CRP has been as a classic indicator of the acute phase response. Thus, until recently, CRP testing has been used primarily for the evaluation of systemic inflammatory diseases including bacterial and viral infections, and in the management of immune disorders such as lupus erythematosis, inflammatory bowel disease, and rheumatoid arthritis.[9]

The initial demonstrations of the prognostic value of CRP in cardiovascular disease were derived from patients with acute coronary ischemia[10] or unstable angina[11] in whom minor elevations of CRP were found to have short-term predictive value. Based in part on these studies, high-sensitivity assays for CRP (hs-CRP) were soon developed to allow for the detection of CRP levels typically within the normal range.[5,12] Recent clinical studies with these hs-CRP tests have confirmed the predictive value of inflammatory testing among patients with non-Q-wave MI,[13,14] severe unstable angina,[15,16] and in the chronic phase after acute infarction.[17] Overall, these data suggest that a heightened inflammatory response is a significant marker of risk for vascular instability and that this enhanced response can be detected with the use of hs-CRP testing.[18]

With regard to primary prevention, the most important data relating hs-CRP and vascular disease derive from large-scale epidemiological studies that have linked minor elevations in CRP among apparently healthy individuals to future risks of MI, stroke, and peripheral vascular occlusion.[19–24] To date, a remarkably consistent series of prospective studies have found that individuals with levels of hs-CRP in the upper quartile have a two- to four-fold increase in risk of future vascular

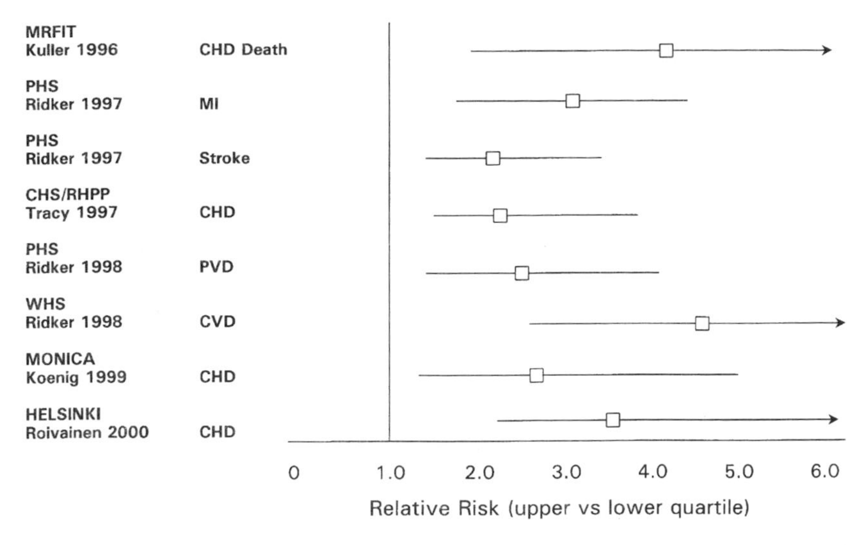

Figure 1. Prospective studies of high-sensitivity C-reactive protein as a risk factor for future cardiovascular events among apparently healthy individuals. CHD = congenital heart disease; MI = myocardial infarction; PVD = pulmonary vascular disease; CVD = cardiovascular disease. From Ridker PM, Haughie P. Prospective studies of C-reactive protein as a risk factor for cardiovascular disease. *J Invest Med* 1998·:391–395.

events compared with those with lower levels of CRP (Fig. 1). This effect is largely independent of other traditional cardiovascular risk factors including hyperlipidemia. In the Physicians Health Study, for example, the risk of future coronary events associated with elevated levels of hs-CRP were independent of total and high-density lipoprotein (HDL) cholesterol, age, smoking status, obesity, hypertension, family history of coronary disease, homocysteine, and fibrinogen.[19] Moreover, in that study, the addition of hs-CRP assessment to total and HDL cholesterol screening significantly improved the ability to predict future risk of both MI and stroke (Fig. 2).[25]

Similar data have been reported in the elderly,[22] in high-risk smokers,[23] in postmenopausal women,[20] and in a European cohort of apparently healthy men.[24] Recent data also indicate that hs-CRP levels appear to predict future arterial occlusions even among individuals with low levels of low-density lipoprotein (LDL) cholesterol. Thus, the detection of an enhanced inflammatory response may help to better predict coronary risk among individuals who suffer acute MI despite having unremarkable lipid levels.

The mechanisms whereby CRP impacts upon vascular risk remain uncertain. Although some data suggest that CRP may have direct vascular effects on thrombosis and cellular adhesion molecule func-

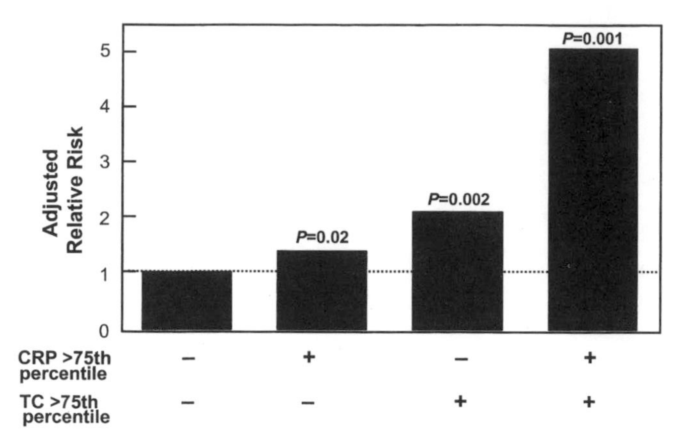

Figure 2. Adjusted relative risks of future myocardial infarction according to baseline levels of total cholesterol (TC) and baseline levels of high-sensitivity C-reactive protein (CRP). From Reference 25.

tion,[26,27] it is probable that elevated levels of CRP are simply an indirect marker of an enhanced state of cytokine-mediated inflammation.[18] In this regard, clinical studies now indicate that several components of the inflammatory system are directly involved in the production of CRP and in the genesis of acute coronary events. For example, elevated levels of IL-6, the primary driver of hepatic CRP production, have been found to predict vascular mortality in the elderly[28] and have been associated with first ever MI among currently healthy middle-aged men.[29] Similarly, among stable post-MI patients, persistent elevation of the inflammatory response as detected by tumor necrosis factor is also associated with increased long-term vascular risk.[30] In the acute coronary syndromes, IL-6 and IL-1Ra have similarly been found to have short-term predictive value.[31]

As an alternative explanation, several investigators have proposed that elevated levels of hs-CRP reflect chronic infection with several different infectious pathogens including *Chlamydia pneumoniae, Helicobacter pylori,* herpes simplex virus, and cytomegalovirus. Although this explanation has broad appeal,[32] it is important to recognize that recent large-scale prospective studies have not found positive associations between immunoglobulin G titers directed against these infectious organisms and measured CRP levels, or between the immunoglobulin G titers and subsequent vascular risk.[33–35] Thus, whether chronic infection has a role in this process remains an open question.

Beyond playing a role as a novel marker for vascular disease, hs-CRP levels may be modifiable as suggested in clinical studies. For example, the magnitude of the cardioprotective effects of low-dose aspirin

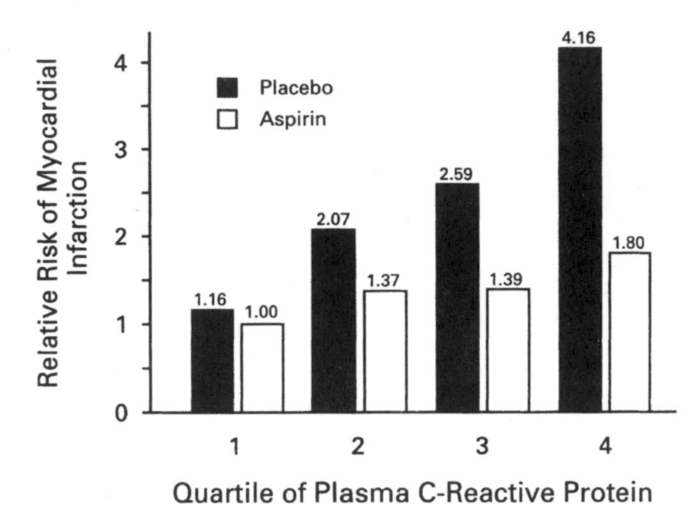

Figure 3. Relative risk of future myocardial infarction associated with baseline levels of high-sensitivity C-reactive protein (hs-CRP), according to aspirin or placebo assignment. The attributable risk reduction for aspirin is 56% among those with high hs-CRP levels but sequentially lower for those with lower levels of hs-CRP. From Reference 20.

observed in the Physicians' Health Study was directly related to the underlying inflammatory response such that those with the highest baseline levels of CRP had the greatest relative benefit (Fig. 3).[19] Thus, it is has been hypothesized that part of the benefit of aspirin may reflect anti-inflammatory as well as antiplatelet properties, data corroborated by experimental work suggesting that hs-CRP levels can be reduced with aspirin consumption.[36]

Experimental and clinical work also suggests that hydroxymethylglutaryl coenzyme A reductase inhibition with statins may have direct anti-inflammatory effects. In experimental models, statin-treated animals have been found to have reduced collagenase expression, increased collagen content within plaques, decreased macrophage content, and an overall decrease in tissue factor expression, all properties potentially beneficial for plaque stabilization.[37–40] This potential has now been directly tested in humans in the Cholesterol and Recurrent Events (CARE) trial where the randomized use of pravastatin was found to greatly attenuate the risk of recurrent coronary events associated with inflammation.[17] Specifically, in the CARE trial, those with elevated levels of hs-CRP at baseline were found to have significantly increased risks of recurrent coronary events ($RR = 1.8$, $P = 0.02$). However, the association between hs-CRP and subsequent risk was significant among those randomized to placebo ($RR = 2.1$, $P = 0.48$), but was

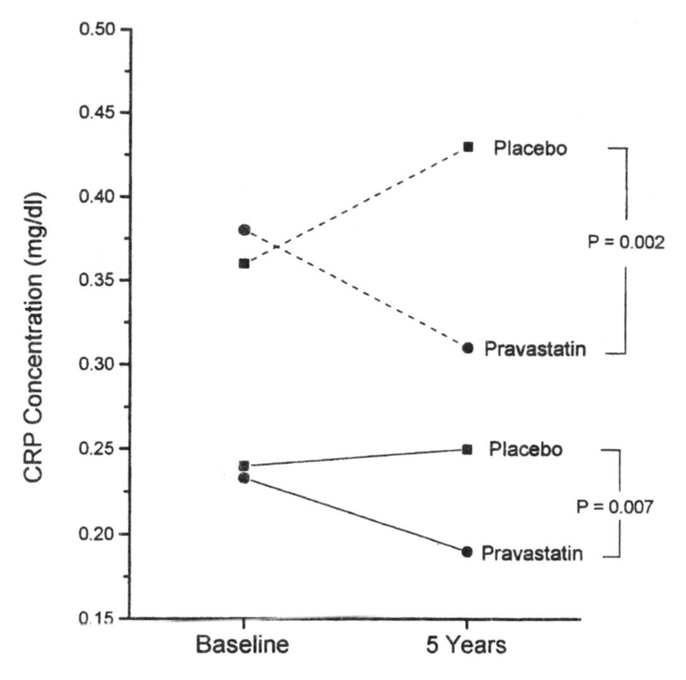

Figure 4. Effects of pravastatin on high-sensitivity C-reactive protein (hs-CRP). Median (solid lines) and mean (dotted lines) are shown for hs-CRP at baseline and at 5 years, according to placebo or pravastatin assignment. From Reference 41.

attenuated and no longer statistically significant among those assigned to pravastatin (RR = 1.29, $P = 0.5$). Moreover, in that trial, the randomized use of pravastatin was associated with significant reductions in plasma levels of hs-CRP, an effect which was independent of any pravastatin-induced changes in LDL cholesterol (Fig. 4).[41] Thus, both clinical and experimental data suggest that statin therapy, in addition to lipid lowering, may have direct effects on the inflammatory response.[42]

Other preventive therapies, including diet and exercise, also appear to interact with the inflammatory response such that those with a sedentary lifestyle or increased body mass index tend to have elevated levels of hs-CRP.[43,44] By contrast, diet and exercise appear to reduce levels of IL-6 and hs-CRP.

Relatively little is known about the effects of other pharmacologic agents on plasma levels of hs-CRP. However, both cross-sectional[45] and interventional studies[46] indicate that the use of estrogen replacement therapy may adversely increase levels of hs-CRP, data which may provide insight into the apparent early hazards associated with hormone replacement therapy observed in the Heart and Estrogen/progestin

Replacement Study (HERS).[47] It is also of interest to note that the effects of estrogen on the acute phase response are not easily predicted. In one study, hs-CRP levels increased with hormone replacement therapy, whereas levels of another acute phase reactant, fibrinogen, decreased.[46] Thus, simple hepatic induction does not seem to explain these results.

Presently, clinical guidelines for hs-CRP testing as an adjunct to lipid screening have not been fully developed. Further, it is important to recognize that the ability of a given parameter to predict risk on a population basis does not necessarily imply that it will accurately predict risk for a given patient. Nonetheless, the United States Food and Drug Administration has recently approved the first commercially based assay for hs-CRP for use in cardiovascular risk prediction. In comparative clinical studies, this high-throughput commercial assay appears to accurately reproduce hs-CRP levels based on previously available research assays (Fig. 5).[5] Further, this commercial assay has

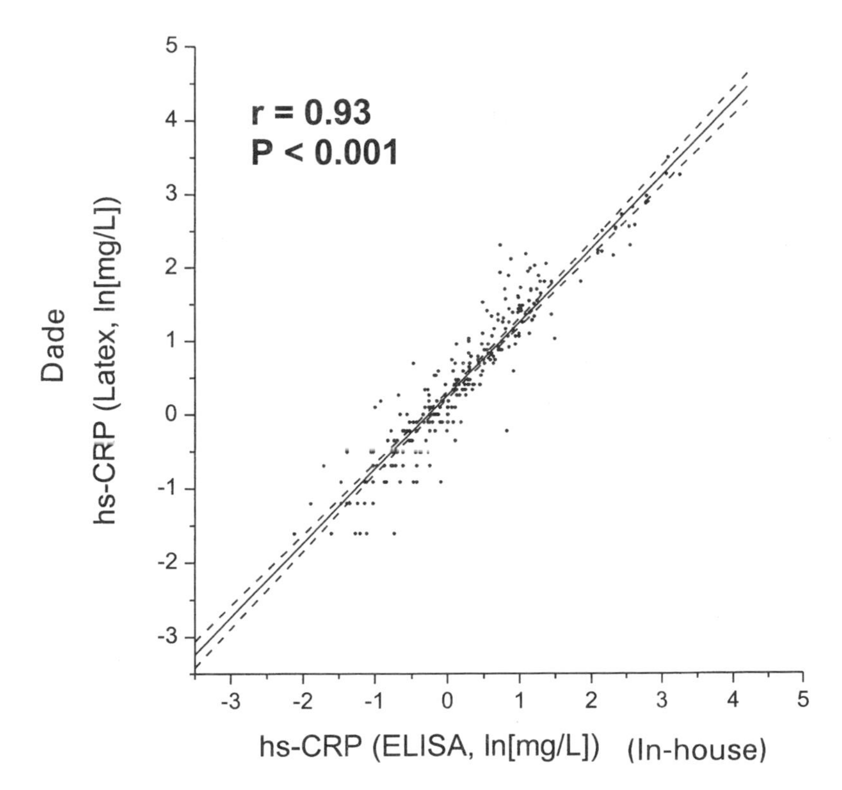

Figure 5. Correlation between in-house enzyme-linked immunosorbent assay (ELISA) for high-sensitivity C-reactive protein (hs-CRP) used in several epidemiological studies and a Food and Drug Administration-approved commercial Latex assay (Dade-Behring, Deerfield, IL). From Reference 5.

been shown to predict cardiovascular events with the same fidelity as research-based assays used in several major prospective clinical studies.[5] Thus, algorithms for risk prediction using hs-CRP must now be developed and carefully tested. Because the dynamic range is wide for all acute phase reactants, it is generally recommended that screening for hs-CRP not be done within a 2-week period of any acute bacterial or viral infection.

Based on the role of inflammation in plaque stability, clinical trials directly targeting the inflammatory response are likely to begin among patients with acute coronary syndromes. Because levels of IL-6,[28,29] sICAM-1,[48,49] and hs-CRP[19–23] have all been shown to be elevated many years in advance of first vascular events, a rationale can also be made to consider targeted anti-inflammatory therapy in the primary as well as secondary prevention of coronary disease. Thus, serum markers of inflammation such as hs-CRP may also be useful along with new therapies as they are developed for the prevention and treatment of cardiovascular disease.

References

1. Ridker PM. Evaluating novel cardiovascular risk factors: Can we better predict heart attacks? *Ann Intern Med* 1999;130:933–937.
2. Lagrand WK, Visser CA, Hermens WT, et al. C-reactive protein as a cardiovascular risk factor: More than an epiphenomenon? *Circulation* 1999;100: 96–102.
3. Libby P. Molecular bases of the acute coronary syndromes. *Circulation* 1995; 91:2844–2850.
4. Ross R. Atherosclerosis—an inflammatory disease. *N Engl J Med* 1999;340: 115–126.
5. Rifai N, Tracy RP, Ridker PM. Clinical efficacy of an automated high-sensitivity C-reactive protein assay. *Clin Chem* 1999;45:2136–2141.
6. Libby P, Ridker PM. Novel inflammatory markers of coronary risk: Theory versus practice. *Circulation* 1999;100:1148–1150.
7. Tillot WS, Francis T. Serological reactions in pneumonia with a non-protein somatic fraction of pneumococcus. *J Exp Med* 1930;52:561–571.
8. Pepys MG. The acute phase response and C-reactive protein. In Weatherall DJ, Ledingham JGG, Warrell DA (eds): *Oxford Textbook of Medicine*. 3rd ed. Oxford, England: Oxford University Press; 1995:1527–1533.
9. Young B, Gleeson M, Cripps A. C reactive protein: A critical review. *Pathology* 1991;23:118–124.
10. Liuzzo G, Biasucci LM, Gallimore JR, et al. The prognostic value of C-reactive protein and serum amyloid A protein in severe unstable angina. *N Engl J Med* 1994;331:417–424.
11. Haverkate F, Thompson SG, Pyke SDM, et al. Production of C-reactive protein and risk of coronary events in stable and unstable angina. *Lancet* 1997;349:462–466.
12. Macy EM, Hayes TE, Tracy RP. Variability in the measurement of C-reactive

protein in healthy subjects: Implications for reference interval and epidemiologic applications. *Clin Chem* 1997;43:52–58.
13. Morrow D, Rifai N, Antman E, et al. C-reactive protein is a potent predictor of mortality independently and in combination with troponin T in acute coronary syndromes. *J Am Coll Cardiol* 1998;31:1460–1465.
14. Toss H, Lindahl B, Siegbahn A, Wallentin L, for the FRISC Study Group. Prognostic influence of fibrinogen and C-reactive protein levels in unstable coronary artery disease. *Circulation* 1997;96:4204–4210.
15. Liuzzo G, Buffon A, Biasucci LM, et al. Enhanced inflammatory response to coronary angioplasty in patients with severe unstable angina. *Circulation* 1998;98:2370–2376.
16. Biasucci LM, Liuzzo G, Grillo RL, et al. Elevated levels of C-reactive protein at discharge in patients with unstable angina predict recurrent instability. *Circulation* 1999;99:855–860.
17. Ridker PM, Rifai N, Pfeffer MA, et al, for the Cholesterol and Recurrent Events (CARE) Investigators. Inflammation, pravastatin, and the risk of coronary events after myocardial infarction in patients with average cholesterol levels. *Circulation* 1998;98:839–844.
18. Maseri A. Inflammation, atherosclerosis, and ischemic events—exploring the hidden side of the moon [editorial]. *N Engl J Med* 1997;336:1014–1016.
19. Ridker PM, Cushman M, Stampfer MJ, et al. Inflammation, aspirin, and the risk of cardiovascular disease in apparently healthy men. *N Engl J Med* 1997;336:973–979.
20. Ridker PM, Buring JE, Shih J, et al. Prospective study of C-reactive protein and the risk of future cardiovascular events among apparently healthy women. *Circulation* 1998;98:731–733.
21. Ridker PM, Cushman M, Stampfer MJ, et al. Plasma concentration of C-reactive protein and risk of developing peripheral vascular disease. *Circulation* 1998;97:425–428.
22. Tracy RP, Lemaitre RN, Psaty BM, et al. Relationship of C-reactive protein to risk of cardiovascular disease in the elderly. Results from the Cardiovascular Health Study and the Rural Health Promotion Project. *Arterioscler Thromb Vasc Biol* 1997;17:1121–1127.
23. Kuller LH, Tracy RP, Shaten J, Meilahn EN, for the MRFIT Research Group. Relationship of C-reactive protein and coronary heart disease in the MRFIT nested case-control study. *Am J Epidemiol* 1996;144:537–547.
24. Koenig W, Sund M, Froelich M, et al. C-reactive protein, a sensitive marker of inflammation, predicts future risk of coronary heart disease in initially healthy middle-aged men: Results from the MONICA (Monitoring trends and determinants in cardiovascular disease) Augsberg Cohort Study, 1984 to 1992. *Circulation* 1999;99:237–242.
25. Ridker PM, Glynn RJ, Hennekens CH. C-reactive protein adds to the predictive value of total and HDL cholesterol in determining risk of first myocardial infarction. *Circulation* 1998;97:2007–2011.
26. Cermak J, Key NS, Bach RR, et al. C-reactive protein induces human peripheral blood monocytes to synthesize tissue factor. *Blood* 1993;82:513–520.
27. Zouki C, Beauchamp M, Baron C, Filep J. Prevention of in vitro neutrophil adhesion to endothelial cells through shedding of L-selectin by C-reactive protein and peptides derived from C-reactive protein. *J Clin Invest* 1997;100:522–529.
28. Harris TB, Ferrucci L, Tracy RP, et al. Associations of elevated interleukin-6 and C-reactive protein levels with mortality in the elderly. *Am J Med* 1999;106:506–512.

29. Ridker PM, Rifai N, Stampfer MJ, Hennekens CH. Plasma concentration of interleukin-6 and the risk of future myocardial infarction among apparently healthy men. *Circulation* 2000;101:1767–1772.
30. Ridker PM, Rifai N, Pfeffer M, et al, for the Cholesterol and Recurrent Events (CARE) Investigators. Elevation of tumor necrosis factor and increased risk of recurrent coronary events following myocardial infarction. *Circulation* 2000;101:2149–2153.
31. Biasucci LM, Liuzzo G, Fantuzzi G, et al. Increasing levels of interleukin (IL)-1Ra and IL-6 during the first 2 days of hospitalization in unstable angina are associated with increased risk of in-hospital coronary events. *Circulation* 1999;99:2079–2084.
32. Danesh J, Collins R, Pete R. Chronic infections and coronary heart disease: Is there a link? *Lancet* 1997;350:430–436.
33. Ridker PM, Kundsin R, Stampfer MJ, et al. Prospective study of *Chlamydia pneumoniae* IgG seropositivity and risks of future myocardial infarction. *Circulation* 1999;99:1161–1164.
34. Ridker PM, Hennekens CH, Stampfer MJ, Wang F. Prospective study of herpes simplex virus, cytomegalovirus, and the risk of future myocardial infarction and stroke. *Circulation* 1998;98:2796–2799.
35. Ridker PM, Hennekens CH, Buring JE, et al. Baseline IgG antibody titers to *Chlamydia pneumoniae, Helicobacter pylori*, herpes simplex virus, and cytomegalovirus and the risk for cardiovascular disease in women. *Ann Intern Med* 1999;131:573–577.
36. Ikonomidis I, Andreotti F, Economou E, et al. Increased proinflammatory cytokines in patients with chronic stable angina and their reduction by aspirin. *Circulation* 1999;100:793–798.
37. Aikawa M, Rabkin E, Okada Y, et al. Lipid lowering by diet reduces matrix metalloproteinase activity and increases collagen content of rabbit atheroma: A potential mechanism of lesion stabilization. *Circulation* 1998;97:2433–2444.
38. Aikawa M, Voglic SJ, Rabkin E, et al. An HMG-CoA reductase inhibitor (cerivastatin) suppresses accumulation of macrophages expressing matrix metalloproteinases and tissue factor in atheroma of WHHL rabbits. *Circulation* 1998;98:I-47. Abstract.
39. Shiomi M, Ito T, Tsukada T, et al. Reduction of serum cholesterol levels alters lesional composition of atherosclerotic plaques: Effect of pravastatin sodium on atherosclerosis in mature WHHL rabbits. *Arterioscler Thromb Vasc Biol* 1995;15:1938–1944.
40. Williams JK, Sukhova GK, Herrington DM, Libby P. Pravastatin has cholesterol-lowering independent effects on the artery wall of atherosclerotic monkeys. *J Am Coll Cardiol* 1998;31:684–691.
41. Ridker PM, Rifai N, Pfeffer M, et al. Long-term effects of pravastatin on plasma concentration of C-reactive protein. *Circulation* 1999;100:230–235.
42. Rosenson RS, Tangney CC. Antiatherothrombotic properties of statins: Implications for cardiovascular event reduction. *JAMA* 1998;279:1643–1650.
43. Smith JK, Dykes R, Douglas JE, et al. Long-term exercise and atherogenic activity of blood mononuclear cells in persons at risk of developing ischemic heart disease. *JAMA* 1999;281:1722–1727.
44. Visser M, Bouter LM, McQuillen GM, et al. Elevated C-reactive protein levels in overweight and obese adults. *JAMA* 1999;282:2131–2135.
45. Ridker PM, Hennekens CH, Rifai N, et al. Hormone replacement therapy and increased plasma concentration of C-reactive protein. *Circulation* 1999;100:713–716.

46. Cushman M, Legault C, Barrett-Connor E, et al. Effect of postmenopausal hormones on inflammation sensitive proteins: The Postmenopausal Estrogen/Progestin Interventions (PEPI) Study. *Circulation* 1999;100:717–722.
47. Hulley S, Grady D, Bush T, et al. Randomized trial of estrogen plus progestin for secondary prevention of coronary heart disease in postmenopausal women. Heart and Estrogen/progestin Replacement Study (HERS) Research Group. *JAMA* 1998;280:605–613.
48. Ridker PM, Hennekens CH, Roitman-Johnson B, et al. Plasma concentration of soluble intercellular adhesion molecule 1 and risks of future myocardial infarction in apparently healthy men. *Lancet* 1998;351:88–92.
49. Hwang S-J, Ballantyne CM, Sharrett AR, et al. Circulating adhesion molecules VCAM-1, ICAM-1, and E-selectin in carotid atherosclerosis and incident coronary heart disease cases. The Atherosclerosis Risk in Communities (ARIC) Study. *Circulation* 1997;96:4219–4225.

Chapter 16

Nuclear Factor Kappa B: *A Marker of Coronary Artery Disease Activity?*

George K. Daniel, MD, Richa Gupta, MD, Jessica Gillespie, BS, Rose Felten, MS, Linda Cise, MS, Kathy Sturdevant, BS, and Michael E. Ritchie, MD

Introduction

Despite substantial progress in the diagnosis and treatment of the manifestations of coronary artery disease, it is currently not possible to identify a priori those specific patients at risk for subsequent sudden cardiac death, myocardial infarction (MI), or unstable angina. Moreover, though specific treatment goals that decrease mortality in patient groups are known, current technology is incapable of assessing the effectiveness of therapy for individual patients.

Current evidence suggests that the unpredictable rupture of coronary artery plaques is due to an acute inflammatory reaction in atherosclerotic vessels and implies that the reduction of events by salicylates and hydroxymethylglutaryl coenzyme A (HMG-CoA) reductase inhibitors are via anti-inflammatory effects.[1–11] Many of the genes involved in the acute inflammatory response, such as those encoding interleukins, tumor necrosis factor, and adhesion molecules, are potently activated by the transcription factor nuclear factor kappa B (NFkB).[12,13] In the presence of anti-inflammatory levels of either acetylated or nonacetylated salicylic acid (ASA and NSA, respectively) in vitro, NFkB can no

This work was supported in part by grants from the NIH (MER), AHA (MER), and VA (MER).

From: Adams JE III, Apple FS, Jaffe AS, Wu AHB (eds). *Markers in Cardiology: Current and Future Clinical Applications.* Armonk, NY: Futura Publishing Company, Inc.; © 2001.

longer act as a transcriptional activator.[14] NFkB activity in atherosclerotic rabbit aortas is also inhibited by HMG-CoA reductase inhibitors.[15] Hence, it was postulated that the activation of NFkB mediates coronary atherosclerotic plaque rupture and that the inhibition of NFkB by anti-inflammatory levels of ASA/NSA or HMG-CoA reductase inhibitors in vivo is the mechanism by which these agents lower clinical events.

Analyses designed to begin to examine these postulates demonstrated that NFkB is selectively and markedly activated acutely in patients with unstable angina and in those with stable angina who rupture coronary plaques within 24 hours of presentation.[16] This study also demonstrated measurable patient-to-patient variability in NFkB activation in those with stable angina. These data indicate that NFkB may be mechanistically involved in plaque rupture, and imply that systemic NFkB activity may be related to coronary artery disease activity. However, further analyses are required to establish a functional role for NFkB in the coronary artery disease process. Accordingly, to begin to determine the mechanistic role of NFkB in plaque rupture and its potential as an indicator of coronary artery disease activity, additional analyses were undertaken.

First, to evaluate the capacity of systemic NFkB levels to serve as an indicator of coronary disease activity, the relationship between NFkB elevation and subsequent coronary events in the first year following the initial evaluation of the 51 stable angina patients in the above-cited study was determined. Second, and as a first step in defining the mechanistic role of NFkB in the development of an acute coronary syndrome, the circulatory location of NFkB activation and its relationship to clinical syndrome were assessed.

Patient Selection and Definitions

All human studies were approved by the Internal Review Boards of the University of Cincinnati in Cincinnati, OH and Indiana University in Indianapolis, IN. Patient selection and preparation of samples at the University of Cincinnati have been published.[16] The patients with stable angina from the published study were used for the 1-year follow-up analyses detailed in this chapter. Indiana University and Roudebush Veterans Affairs Medical Center patients undergoing coronary angiography for evaluation of stable or unstable angina classified as stable or unstable angina Class I, II, or III according to Braunwald's classification were eligible for the portion of this study designed to identify the systemic versus coronary artery circulatory location of NFkB activation.[17] Patients with recent or ongoing MI were excluded from both portions of the study. Clinical characteristics that were evaluated included age, presence or absence of hypertension, diabetes mellitus, dyslipidemia, history of smoking, severity of coronary artery disease (defined as epicardial coronary artery stenosis of at least 70% in one or more coronary arteries), and left ventricular function. Use of aspirin, HMG-CoA reductase inhibitors, and other lipid-lowering agents, angiotensin-converting enzyme inhibitors, and β-blockers were also tracked.

Blood Sample Collection and Electromobility Shift Assays

Samples from patients participating in the original study were isolated from arterial or venous sites, or both. The preparation of samples for determination and quantification of NFkB have been described.[16] For the current study, results of this single-point-in-time NFkB level were evaluated in the context of the patient's subsequent events over the ensuing 1 year. For determining the location of NFkB activation, blood samples were collected from the ascending aorta (AA), coronary sinus (CS), superior vena cava (SVC), inferior vena cava (IVC), and coronary arteries (left and right) of 21 patients undergoing cardiac catheterization for evaluation of stable or unstable angina. Only data derived from non-coronary-artery samples were evaluated for this study. Samples were coded, labeled, and sent to the laboratory, where isolation of nuclear proteins and electromobility shift assays were performed by research personnel unaware of the patient's clinical condition as described.[16]

One-Year Follow-up Analyses: The Level of NFkB Activation Correlates with the Likelihood and Acuity of a Subsequent Event

To evaluate the relationship between systemic NFkB activity and the likelihood and temporal acuity of a subsequent coronary event, 1-year follow-up analyses were performed on those with stable angina enrolled in the original trial. The initial demographics of the 51 patients with stable angina were detailed previously.[16] One-year follow-up analyses showed no significant increase in the development of clinical conditions associated with increased cardiac risk. Though there was an increase in the use of medications known to lower clinical cardiac events, there was no relationship between the lack of initiation of any of these medications and an earlier clinical event.

Nineteen (19.6%) of 51 patients had a clinical event as defined by death, coronary artery bypass grafting (CABG), hospitalization for a documented coronary artery syndrome (either unstable angina defined as chest pain accompanied by diagnostic electrocardiographic changes or diagnostic increases in troponin or creatine kinase or MI), or stroke. There were four (7.5%) deaths, and seven patients (13%) underwent CABG. One death occurred at the end of 1 year as a result of a stroke (2%), and one death occurred after CABG. Two deaths from complicated acute MI occurred within 2 months of initial NFkB analysis. Both of these early deaths were associated with NFkB levels greater than 1.5 AU. There were 16 acute coronary syndrome episodes in 13 (25%) patients. In 11 of the 13, NFkB levels were greater than 1 AU. All those having an acute coronary syndrome event within 2 months of NFkB assessment (8 episodes in 8 patients) had NFkB levels greater than 1.5 AU. Twenty-eight patients did not meet a clinical endpoint, with the vast majority (25 of 28) having an initial NFkB level less than 1 AU. Taken together, these data suggest that incremental increases in systemic NFkB activation are associated with the more rapid onset of clinical cardiac events, implying that NFkB levels may be predictive of events in individual patients.

Circulatory Location of NFkB Activation: NFkB Activation Across Coronary Bed

To determine the circulatory location of NFkB activation, a consecutive series of patients undergoing coronary angiography for evaluation of stable or unstable coronary syndromes were evaluated. Evidence of NFkB activation was determined in white blood cells isolated from the AA, CS, and SVC. Of the 21 patients enrolled, 13 had stable angina and 8 had unstable angina. Evidence of NFkB activation across the coronary bed defined by a CS:AA ratio of greater than 1 was observed in 10 of 13 patients with stable angina (mean 2.19 ± 0.85) but in only 2 of 8 patients with unstable angina (mean 0.85 ± 0.73). This difference was statistically significant ($P = 0.005$), demonstrating a relationship between coronary bed NFkB activation and clinical syndrome.

Circulatory Location of NFkB Activation: Systemic versus Coronary Circulation

To compare the degree of NFkB activation across the coronary bed with systemic NFkB activation, NFkB activation was assessed in the

CS and compared with that in the SVC. When evaluated in the context of the patient's presenting complaint, these analyses also allowed for an investigation into the relationship between systemic and coronary bed levels of NFkB activation and clinical syndrome. Consistent with previous work, systemic NFkB levels were increased in all patients with unstable angina. In 7 of 8 with unstable angina, systemic NFkB activation was greater than coronary bed NFkB activation (SVC:CS >1; mean 2.32 ± 1.36). The SVC:CS ratio was greater than 1 in only 4 of 13 patients (mean 0.97 ± 0.62) with stable angina. The difference in SVC: CS ratio of NFkB activation between the two clinical syndromes was statistically significant ($P = 0.014$).

The SVC:CS ratio of greater than 1 in unstable angina was a result of both increased SVC NFkB levels of activation and decreased CS levels of activation. It is important to note that the level of NFkB activation in the CS of those with unstable angina was also lower than that observed in the CS effluent of those with stable angina. This reduced level of NFkB activation across the coronary bed in unstable patients was not due to a lack of propensity for activation, as all CS effluents retained the capacity to be appropriately stimulated ex vivo with lipopolysaccharide.[17]

An additional observation that can be derived from these data is that as patients became more unstable, CS levels of NFkB activation began to fall and SVC levels of NFkB activation started to rise. These changes account for the variability among the patients and the lack of 100% consistency between coronary bed or systemic NFkB activation and clinical syndrome.

Discussion

We have previously demonstrated a relationship between systemic levels of NFkB activation and unstable angina.[16] Because two stable patients with markedly increased systemic NFkB levels became unstable within 24 hours of presentation, these initial analyses also suggested that activation of NFkB might predate the acute clinical event, thus implicating NFkB mechanistically. To begin to establish a mechanistic role for NFkB in the process and progression of coronary artery disease and to start to evaluate its candidacy as an indicator of disease activity and a target for therapy, the relationship between a single systemic NFkB level and the likelihood of a subsequent clinical event was determined. As our data illustrate, in patients with stable angina pectoris there is a remarkable relationship between an increase in systemic NFkB activation and the temporal acuity of a subsequent clinical event. Such results strongly suggest that systemic activation of NFkB precede and may be mechanistically involved in acute plaque rupture and support the conjecture that systemic NFkB levels indicate coronary artery disease activity.

Insight into the mechanistic role of NFkB in coronary artery disease progression can be gained by localizing the site of NFkB activation and correlating it with clinical syndrome. Our initial expectation was that progression in clinical instability would result in increases in both systemic (SVC:CS) and coronary bed (CS:AA) activation. In other words, patients with unstable angina were anticipated to have increased levels of systemic NFkB activity accompanied by increases in NFkB activation through the coronary bed as manifested by increased NFkB activity in the CS effluent. Those with stable coronary syndromes were expected to have low levels of systemic NFkB activity with little NFkB activation occurring through the coronary bed. As predicted and consistent with the initial study, unstable coronary syndromes were associated with increased systemic NFkB activation, and stable coronary syndromes had low but variable levels of systemic NFkB activity. However, analyses investigating NFkB activation across the coronary bed in those with stable versus unstable syndromes provided surprising results. Patients with stable angina, rather than having little NFkB activation across the coronary bed, showed consistent and marked activation of NFkB during transit of the coronary circulation (increased CS:AA ratio). Conversely, patients with unstable angina demonstrated a reduction in NFkB activity consequent of transit through the coronary bed as manifested by marked declines in CS-derived NFkB activity.

Though the cause of the reduction in CS-derived NFkB activity was not determined, there are a number of potential explanations. Cells with activated NFkB may be actively destroyed or turned off by events in the coronary circulation. Alternatively, cells with activated NFkB may be selectively removed from the coronary circulation either by the ruptured plaque or as a consequence of events in the coronary bed activated in response to plaque rupture. Additional experiments are under way to determine the mechanism of this unexpected but exciting observation. We anticipate that these analyses will demonstrate that cells containing activated NFkB are selectively removed from the coronary circulation by the disrupted plaque.

Since it was recognized that inflammation plays a major role in coronary artery disease progression, a number of markers of inflammation have been identified as being related to the presence and extent of coronary artery disease. For example, many studies show a significant relationship between increases in systemic C-reactive protein (CRP) and coronary artery disease.[18,19] Increased CRP is associated with a worse outcome in patients presenting with acute coronary syndromes.[20] Incremental increases in CRP in nonacute patients are also independently associated with incremental increases in the risk of subsequent coronary artery events.[21] As such, increased CRP serves as the equivalent of a coronary artery disease risk factor, much like hypertension, diabetes mellitus, hyperlipidemia, and so on. Although such data strongly support a role for the use of CRP as a measure of coronary artery disease risk, CRP is limited in its ability to function as an indicator of coronary artery disease activity as it relates to the likelihood of an event occurring acutely. The relationship between NFkB activation and coronary artery events presented in this chapter bears a striking relationship to that observed with CRP. However, there are several critically important differences. Perhaps most importantly, we show evidence of active regulation of NFkB by the coronary bed, implicating it mechanistically in the process of coronary artery disease progression. We further demonstrate that rises in NFkB, as opposed to CRP, are directly related to the earlier occurrence of a cardiac event. These data suggest that whereas the primary role of CRP is as a cardiac risk factor, the role of NFkB may be as an indicator of coronary artery disease activity. As such, NFkB could further function as a measure of therapeutic effectiveness. Consequently, additional analyses are under way to test these possibilities.

References

1. Ross R. The pathogenesis of atherosclerosis: A perspective for the 1990s. *Nature* 1993;362;29:801–808.
2. Jonasson L, Holm J, Skalli O, et al. Regional accumulation of T cells, macrophages, and smooth muscle cells in the human atherosclerotic plaque. *Arteriosclerosis* 1986;6:131–138.
3. Sato T, Takebayashi S, Kohehi K. Increased subendothelial infiltration of the coronary arteries with monocytes/macrophages in patients with unstable angina. *Atherosclerosis* 1995;68:191–197.
4. Mazzone A, De Servi S, Ricevuti G, et al. Increased expression of neutrophil and monocyte adhesion molecules in unstable coronary artery disease. *Circulation* 1993;88:358–363.
5. Quinn MT, Parthasarathy S, Fong LG, Steinberg D. Oxidatively modified low density lipoproteins: A potential role in recruitment and retention of monocyte/macrophages during atherogenesis. *Proc Natl Acad Sci U S A* 1987;84:2995–2998.
6. van der Wal AC, Becer AE, van der Loos CM, Das PK. Site of intimal rupture or erosion of thrombosed coronary atherosclerotic plaques is characterized by an inflammatory process irrespective of the dominant plaque morphology. *Circulation* 1994;89:36–44.
7. Harrison DG, Armstrong ML, Freiman PC, Heistad DDL. Restoration of endothelium-dependent relaxation by dietary treatment of atherosclerosis. *J Clin Invest* 1987;80:1808–1811.
8. Sacks FM, Pfeffer MA, Moye LA, et al. The effect of pravastatin on coronary events after myocardial infarction in patients with average cholesterol levels. *N Engl J Med* 1996;335:1001–1009.

9. Cairns A, Gent M, Singer J, et al. Aspirin, sulfinpyrazone, or both in unstable angina: Results of a Canadian Multicenter Trial. *N Engl J Med* 1985;313: 1369–1375.
10. Bataille R, Klein B. C-reactive protein levels as a direct indicator of interleukin-6 levels in humans in vivo. *Arthritis Rheum* 1992;35:982–984.
11. Biasucci LM, Vitelli A, Liuzzo G, et al. Elevated levels of interleukin-6 in unstable angina. *Circulation* 1996;94:874–877.
12. Barnes PJ, Karin M. Mechanisms of disease: Nuclear factor-kB—a pivotal transcription factor in chronic inflammatory diseases. *N Engl J Med* 1997; 336:1041–1045.
13. Lenardo MJ, Baltimore D. NF-kB: A pleiotropic mediator of inducible and tissue-specific gene control. *Cell* 1989;58:227–229.
14. Kopp E, Ghosh S. Inhibition of NFkB by sodium salicylate and aspirin. *Science* 1994;265:956–959.
15. Bustos C, Hernandez-Presa MA, Ortego M, et al. HMG-CoA reductase inhibition by atorvastatin reduces neointimal inflammation in a rabbit model of atherosclerosis. *J Am Coll Cardiol* 1998;32:2057–2064.
16. Ritchie ME. Nuclear factor-KB is selectively and markedly activated in humans with unstable angina. *Circulation* 1998;98:1707–1713.
17. Braunwald E (ed): *Heart Disease: A Textbook of Cardiovascular Medicine*. Philadelphia: W.B. Saunders; 1997:1331–1332.
18. Ridker PM, Cushman M, Stampfer MJ, et al. Inflammation, aspirin, and the risk of cardiovascular disease in apparently healthy men. *N Engl J Med* 1997;336:973–979.
19. Kuller LH, Tracy RP, Shaten J, Meilahn EN. Relationship of C-reactive protein and coronary heart disease in the MRFIT nested case-control study. *Am J Epidemiol* 1996;144:537–547.
20. Liuzzo G, Biasucci LM, Gallimore JR, et al. The prognostic value of C-reactive protein and serum amyloid A protein in severe unstable angina. *N Engl J Med* 1994;331:417–424.
21. Ridker PM, Hennekens CH, Buring JE, Rifai N. C-reactive protein and other markers of inflammation in the prediction of cardiovascular disease in women. *N Engl J Med* 2000;342:836–843.

Chapter 17

The Use of Troponins to Detect Cardiac Injury after Cardiac and Noncardiac Surgery

Jesse E. Adams, III, MD

Several preceding chapters have discussed the challenges of detecting episodes of myocardial ischemia and necrosis in individuals who present with chest discomfort possibly due to an acute coronary syndrome. Although this is a challenge in patients who present with a primary complaint of chest discomfort, the difficulty in detecting potential myocardial ischemia or necrosis is far greater in individuals who are undergoing an operative procedure. The potential for myocardial ischemia must often be considered in the perioperative setting because episodes of hypotension, tachycardia, or hypoxia are not uncommon in diverse surgical situations. The potential for a cardiac cause in such situations is often greatest for patients who have a known history of cardiac disease, but is a legitimate concern in all patients. However, most patients do not experience perioperative myocardial ischemia or necrosis, and thus exclusion of myocardial ischemia is often the most important clinical endpoint.

Diagnostic aids that are routinely used for patients with acute coronary syndromes may be less helpful in individuals who have just had surgery. The history is often of little benefit, although a careful history is still necessary. Patients often are not able to adequately or accurately report if they are experiencing angina symptoms, and episodes of ischemia are often asymptomatic. Additionally, many types of surgery, especially thoracic surgery, require incisions that will result in chest discomfort. To assist in the detection of episodes of ischemia, electrocardiograms (ECGs) are obtained whenever symptoms merit consideration of myocardial ischemia; ECGs are routinely obtained in select patients for surveillance after surgery. As discussed in chapter 6, ECGs

From: Adams JE III, Apple FS, Jaffe AS, Wu AHB (eds). *Markers in Cardiology: Current and Future Clinical Applications.* Armonk, NY: Futura Publishing Company, Inc.; © 2001.

often do not demonstrate diagnostic changes in the setting of potential ischemia. This is especially true in the postoperative setting, when non-diagnostic alterations in the ECG tracing are common. Accordingly, there is increasing interest and reliance on blood-based markers for the detection of myocardial injury in patients after surgery.

Measurement of the MB isoenzyme of creatine kinase (CK-MB) has been used for more than 20 years for the diagnosis of myocardial necrosis. Although there are some limitations of CK-MB in terms of both sensitivity and specificity, these issues are of particular concern for the use of this marker in patients after surgery.[1] CK-MB is present in low amounts in skeletal muscle (1% to 5%) and can occur in significantly increased amounts in individuals with acute or chronic muscle disease or renal insufficiency. Cellular injury to skeletal muscle is associated with many operative procedures. This cellular injury will result in the release of skeletal muscle cell contents and will therefore often result in elevated levels of total CK and CK-MB. Use of an absolute cut-off for CK-MB (as is often used in patients with acute coronary syndromes) in the perioperative period will result in impaired specificity. Although calculating the ratio of CK-MB to total CK has been tried in the past, this strategy also results in unacceptable problems in both sensitivity and specificity.[2] Because of these limitations, there has been increased interest in the use of measurements of levels of troponin in the postoperative period.

Noncardiac Surgery

A large body of literature has developed regarding the occurrence of myocardial injury in the perioperative period in patients who have had noncardiac surgery. Significant gains have been achieved in the appropriate risk prognostication, intraoperative management, and postoperative treatment to ameliorate the occurrence of cardiovascular morbidity and mortality. The likelihood of a perioperative cardiac complication is influenced by the nature of the operative procedure, the hemodynamic stability during the procedure, alterations in myocardial oxygen supply and demand, and the pretest likelihood of coronary artery disease. Studies have shown that postoperative myocardial infarctions (MIs) are most likely after surgical procedures that confer a significant vascular or hemodynamic impact, or both. It has also been found that the incidence of postoperative cardiac complications is not limited to the immediate time of the procedure itself, but extends for up to 3 days. The detection of cardiac injury in the perioperative period, as discussed above, can be problematic, and this has confounded both the clinician and the clinical researcher.

At this time, the optimum method for the detection of perioperative MIs after noncardiac surgery appears to be the serial measurement of levels of cardiac troponin. As elaborated on elsewhere in this monograph, the measurement of levels of troponins has been found to yield powerful diagnostic and prognostic information in patients with cardiac disease. The robust specificity of troponins is critical to their success; the long diagnostic window (5 to 7 days for cardiac troponin I [cTnI] and 5 to 10 days for cardiac troponin T [cTnT]) facilitates their clinical application. With the current-generation assays, measurement of either cTnI or cTnT should provide similar information on the incidence of perioperative MI in patients who have had noncardiac surgery.[2–4] Measurement of cTnI and cTnT appears to provide similar sensitivity and specificity in this population. This is in contrast to earlier studies that suggested a benefit of using cTnI over cTnT. Initial trials of cTnT used the first-generation assay, which used affinity-purified cardiospecific polyclonal capture antibodies that had a low level of reactivity (1% to 3%) with the skeletal muscle form of troponin T.[5] Although this assay format generally provided sufficient specificity for patients with acute coronary syndromes, it could result in problems with specificity in patients with significant muscle injury. Subsequent assay formats of cTnT, which use a dual monoclonal assay format, have resolved this difficulty, and problems of specificity with the current-generation assay format do not appear to be an issue. [6]

Serial measurements of troponins can be used for detecting perioperative MIs in individuals who have had noncardiac surgery, and the reference levels routinely used for patients in other clinical situations (such as those with chest pain) will also apply. As discussed later in this section, this is not the case with the diagnosis of perioperative MI after *cardiac* surgery. Because MIs can occur for up to 3 to 4 days after surgery, the routine serial measurement of cardiac troponins after surgery is recommended, especially for patients felt to be at increased risk for cardiac complications. Since patients often will not manifest symptoms, it would not be advisable to obtain serial measurement of troponins for patients who complain only of chest discomfort. Additional samples can be obtained when any symptoms or abnormalities are observed (such as abnormalities seen on monitoring systems, etc.). It must be remembered, however, that it can take 6 hours or more before one can reliably expect to find the appearance of troponins in circulation after myocardial cell death, and thus one must allow a sufficient time to elapse before sample acquisition. As with patients with acute coronary syndromes, any amount detected (even in the normal range) should be cause for concern and subsequent testing. Any elevation above the reference level should be interpreted as myocardial injury, unless one can demonstrate that analytic errors were to blame.

The measurement of serial levels of troponins can also be used to refine our treatment paradigms in patients after noncardiac surgery to further decrease the likelihood of cardiac morbidity and mortality in the perioperative period.

Cardiac Surgery

The detection of myocardial injury after cardiac surgery is more difficult but is of considerable importance. More than 500,000 patients in the United States, Canada, and Europe undergo coronary artery bypass grafting each year. The annual cost has been estimated to be at least $15 billion per year.[7] It has been estimated that cardiac morbidity and mortality may occur in up to 20% of patients who undergo this procedure. Cardiac morbidity and mortality can occur secondary to acute or chronic ventricular dysfunction, acute or preexisting coronary artery disease that was not remedied by the surgery, incomplete coronary revascularization, MI, renal insufficiency, cerebrovascular events, infections, and hemodynamically significant dysrhythmias. Cardiac injury can occur due to inadequate myocardial protection during aortic cross-clamping, to an unduly long period of time on the heart-lung machine (during which time the myocardium is not perfused), to post-bypass reperfusion injury, to spasm of the internal mammary artery, to direct injury during surgical manipulation, and to postoperative injury or occlusion. Cardiopulmonary bypass has been found to stimulate leukocyte-platelet adhesion, as well as to stimulate the release of proinflammatory molecules such as endotoxin (which among other actions stimulates the complement system) and the proinflammatory mediator tumor necrosis factor.[8,9]

Thus, detection of cardiac injury after cardiac surgery is conceptually different from diagnostic strategies in patients with acute coronary artery syndromes or after noncardiac surgery. In other clinical situations, physicians can theoretically separate patients into two groups: those who have had and those who have not had *any* degree of myocardial cellular necrosis. In the setting of perioperative MI that occurs after cardiac surgery, however, the situation is quite different. As discussed earlier, *every* patient who undergoes cardiac bypass will have some degree of myocardial cell death. The issue becomes one of discriminating those patients who have had a larger amount of myocardial cellular injury (which can potentially affect patient care) from those with less. The question will be whether a "clinically significant" degree of myocardial cell injury has occurred. The reference levels used to detect perioperative myocardial injury will be substantially higher than those thresholds used in other clinical situations and must be defined for this particular post-cardiac-surgery population.

Because of differences among assays (which can be a problem with many analytes but is currently a particular issue with cTnI, as discussed elsewhere in this monograph), at this time studies will result in recommendations that will be assay-dependent. There will also be differences in the recommended levels predicated on the clinical endpoint used to define "abnormal." Additionally, when using or evaluating biomarkers in the post-cardiac-surgery population, it is important to define the time from cardiac surgery (usually defined from the aortic cross-clamp time, which generally corresponds to the onset of the cessation of cardiac circulation) when proposing abnormal reference values. The levels of cardiac troponins and other markers tend to rise and fall rapidly in the first several 24 hours after cardiovascular surgery, and thus proposed reference values must be reported in relation to the time of aortic cross-clamping.

The clinical detection of cardiac injury in the post-cardiac-surgery patient is difficult; symptoms of chest discomfort are largely worthless, dyspnea is common, ECGs frequently manifest nonspecific ST or T-wave changes, and echocardiograms have suboptimal visualization in many patients. This diagnostic difficulty has not only complicated routine patient care. In fact, studies attempting to define the incidence of post-cardiac-surgery MIs have reported rates of 5% to 80%, with mortality rates ranging from 0.5% to 14%.[10] This wide range is due to differing criteria for MI. Use of any elevation of CK-MB as the definition of an infarction, for example, will result in a relatively high incidence of perioperative cardiac complications, whereas reliance on the development of new Q waves postoperatively will result in a much smaller number but will miss the non-Q-wave infarctions. This disagreement in the diagnostic criteria for perioperative MI complicates research in the field. It is difficult, for example, to assess whether a novel cardioplegic agent decreases the incidence of perioperative MI if there is difficulty in the reliable detection of cardiac injury.

There has been great interest in the use of the measurement of cardiac troponins (both I and T) to assist in the detection of cardiac injury after cardiothoracic surgery, due in large part to superior cardiac specificity. Mair et al.[11] first reported on the use of cTnT in 1991 for the detection of perioperative MI in 21 patients undergoing coronary artery bypass grafting. MI was defined as an abnormal ECG associated with increases in CK-MB lasting greater than 12 to 18 hours. Fourteen patients did not meet these criteria, and levels of cTnT did not exceed 1.6 μg/L (upper limit of normal 0.1 μg/L). Two individuals had intraoperative MIs as defined above; all had marked elevations of cTnT, and one of these patients died from cardiogenic shock. The authors proposed that a peak concentration of cTnT less than 2.5 μg/L could exclude clinically relevant cardiac injury. This report was followed by reports by Triggiani et al.[12] and Machler et al.[13] substantiating that the measurement of cTnT could be used to exclude perioperative myocardial infarction with a reference level of 2.5 to 3.0 μg/L. Additionally, the data suggested that patients who had elevations of troponin T before cardiac surgery had a worse prognosis, consistent with previous studies that found that individuals who undergo surgery shortly after MI are at increased perioperative risk.

Mair and colleagues[14] also were among the first to report in 1994 the use of measurements of cTnI in 28 individuals who underwent coronary artery bypass grafting. Cardiac TnI was measured with an assay developed by Larue et al.[15] MI was diagnosed either by the appearance of new persistent Q waves or by CK-MB activity greater than 50 IU/L on the first postoperative day in conjunction with new persistent wall motion abnormalities. Non-Q-wave MI was diagnosed in this study by new ST-segment deviation in conjunction with CK-MB activity greater than 20 IU on the first postoperative day. Levels of cTnI up to 2 μg/L (upper reference limit = 0.1 μg/L) were routinely found in patients without complications. Patients with perioperative MIs (n=6) had peak cTnI concentrations greater than 4.5 μg/L and had peaks occurring approximately 24 hours after aortic cross-clamping.

Etievent et al.[16] used the same assay developed by Larue's group and found that peak cTnI levels occurred at approximately 6 hours after surgery and were usually absent by day 5 in patients who did not have perioperative MIs. A correlation was present between total aortic cross-clamp time and levels of cTnI. Peak levels for patients who did not have perioperative myocardial injury were reported: 1.1 μg/L (normal $<$0.1 μg/L) in patients who had valve replacement and 1.9 μg/L after coronary bypass graft surgery.

Hirsch et al.[17] measured levels of cTnI using the assay developed by Bodor et al.[18] (upper limits of normal 1.5 μg/L) in pediatric cardiac surgery patients after the repair of congenital cardiac lesions, and found a uniform pattern of a postoperative rise and fall in the levels of cardiac troponin in all children who had elevated levels. A correlation was present between aortic cross-clamp time and total cardiac bypass time and levels of cTnI. Levels of cTnI at 12 and 24 hours correlated with outcome (intraoperative support, duration of endotracheal intubation, hospital stay, and the duration of the stay in the intensive care unit [ICU]). A similar conclusion was reached by a more recent study involving open heart surgery in children and infants; levels of cTnI obtained 4 hours after initial admission to the ICU correlated strongly with the need for inotropic support, the severity of renal dysfunction, and the duration of intubation.[19]

Recently, we completed a study of patients who underwent cardiac surgery; initial data from this study have been published.[20] We enrolled

124 patients; the gold standard for the presence of a "significant" perioperative MI was the detection of a new wall motion abnormality on serial echocardiograms. Serial measurements of cTnI were taken using the assay developed by Bodor et al and currently marketed by Dade Behring (Deerfield, IL).[18] Patients who developed new wall motion abnormalities indicative of perioperative myocardial injury had significantly higher levels of cTnI (21 ± 12.88 μg/L) than those who did not (6.30 ± 3.55 μg/L). A cut-off level of 11 μg/L had a negative predictive value of 97%. It is anticipated that the levels of cTnI in this situation would most likely be used to exclude a perioperative MI, thus sparing many patients additional unnecessary evaluation. Patients who had markers suggestive of perioperative myocardial injury would likely have further diagnostic testing (such as echocardiography) done. Accordingly, the use of a reference level of 11 μg/L would be recommended based on these data, whereas the "usual" reference level in patients with acute coronary artery syndromes is 1.5 μg/L according to the package insert. Finally, it has recently been reported that high elevations of cTnI (>60 μg/L) immediately after open heart surgery predicts a worse outcome.[21]

Thus, it appears that levels of troponins can successfully detect cardiac injury after cardiac surgery, and elevations of cardiac troponins may be predictive of postoperative prognosis, although much more work must be done before this is defined. Additionally, most would accept the contention that myocardial protection during cardiac surgery is important, and that lesser amounts of myocardial cell death would be preferred.

Accordingly, measurements of levels of cardiac troponins have been used to assess therapies provided at the time of coronary artery bypass grafting. Hannes et al.[22] used serial measurements of cTnT to assess the use of diltiazem to prevent spasm of internal mammary artery grafts. Patients who received intravenous diltiazem had lower peak levels of cTnT. Krejca et al.[23] used serial measurements of cTnT to assess the application of intermittent aortic cross-clamp, on bypass pump and off-pump with a beating heart. They found that patients who had coronary artery bypass grafting without aortic cross-clamping and without cardiopulmonary bypass had lower elevations of cTnT, which they believed indicated superior myocardial protection. Wendel et al.[24] have used measurements of cTnT to assess the administration of aprotinin during aortic cross-clamping. Pelletier et al.[25] used levels of cTnT in 120 patients undergoing cardiac surgery to compare the myocardial protection provided by intermittent antegrade warm versus cold blood cardioplegia. It has been suggested that a number of cytokine pathways are activated during cardiopulmonary bypass, especially those of inflammatory cytokines. Recently, it has been noted that levels of cTnI correlate with levels of interleukin 8 after cardiopulmonary bypass, and that the use of heparin-coated tubing in the bypass circuits resulted in decreased levels of both interleukin 8 and cTnI, consistent with a lesser degree of perioperative cardiac injury.[26]

A major problem in the application of troponin measurements, however, is the difficulty of comparing results from different assays as well as the results from different-generation assays from the same manufacturer. This is particularly a problem with assays for cTnI; measurement of cTnT has largely been spared this issue due to the availability of only one commercially available assay format. It has been found that uncorrected values for levels of cTnI can vary up to 60-fold.[20,27] This disparity precludes direct comparison of the levels of troponin between different assay formats. Although this can cause difficulties in many clinical and research settings, it is particularly an issue in the detection of perioperative MIs. A gold standard is difficult in these situations; thus, clinical studies must be tied to robust and clearly defined clinical endpoints. However, any trial will ultimately result in a recommended troponin level that correlates with some clinical endpoints (such as mortality or a new abnormality on echocardiogram), but the reported level of troponin will only be of use to those clinicians to whom that particular troponin assay is available. At this time, it is impossible to make a recommendation of using a cut-off of 11 μg/L for a particular assay, and be able to "convert" that recommendation to any other assay format for troponin I or to cTnT. Much work, by the manufacturers as well as by the research community, remains to be done to resolve this lack of standardization of troponin assay systems.[28]

The measurement of serial troponins can be used to detect cardiac injury in the postoperative patient. In patients who have had noncardiac surgery, the tests should be used as they are in other clinical situations, that is, the routine reference cut-offs apply. In applying measurements of troponins to patients after cardiac surgery, however, all patients have some degree of cardiac cell death, and thus reference levels will be higher and must be defined for this specific patient population. Unfortunately, these recommendations will be assay-dependent, and currently this is especially a problem with cTnI. At present, we use the measurement of cTnI in our post-open-heart patients as a screening test. Those patients who have no elevation of troponin I above 11 μg/L (using the Dade assay) 8 to 16 hours after aortic cross-clamping do not require further evaluation for significant perioperative myocardial injury. For those patients with troponins greater than 11 μg/L, however, we commonly proceed with further diagnostic testing (most commonly echocardiography) to provide additional information. It must be recognized that elevations of troponins do not provide an etiologic cause but simply demonstrate the presence of myocardial cellular necrosis. Necrosis in the perioperative period can occur by a variety of mechanisms such as poor myocardial protection during cardioplegia or occlusion of a bypass graft that will have different therapeutic implications. Further studies are necessary to refine our application of the measurement of troponin proteins to this challenging population.

References

1. Adams JE, Abendschein DR, Jaffe AJ. Biochemical markers of myocardial injury: Is MB creatine kinase the choice of the 1990s? *Circulation* 1993;88: 750–763.
2. Adams JE, Sicard GA, Allen BT, et al. Diagnosis of perioperative myocardial infarction with measurement of cardiac troponin I. *N Engl J Med* 1994;330: 670–674.
3. Badner NH, Knill RL, Brown JE, et al. Myocardial infarction after noncardiac surgery. *Anesthesiology* 1998;88:572–578.
4. Poldermans D, Cobbaert C, Struijik L, et al. Diagnostic utility of troponin T to exclude perioperative cardiac events in patients undergoing major vascular non-cardiac surgery. *J Am Coll Cardiol* 1998;32:342A.
5. Katus KA, Remppis A, Scheffold T, et al. Intracellular compartmentation of cardiac troponin T and its release kinetics in patients with reperfused and nonreperfused myocardial infarction. *Am J Cardiol* 1991;67:1360–1367.
6. Wu AHB, Valdes R, Apple FS, et al. Cardiac troponin-T immunoassay for diagnosis of acute myocardial infarction. *Clin Chem* 1994;40:900–907.
7. Mangano DT. Perioperative cardiac morbidity-epidemiology, costs, problems, and solutions [editorial]. *West Med J* 1994;161:87–89.
8. Jansen NJG, van Oeveren W, Gu YJ, et al. Endotoxin release and tumor necrosis factor formation during cardiopulmonary bypass. *Ann Thorac Surg* 1992;54:744–748.
9. Rinder CS, Bonan JL, Rinder HM, et al. Cardiopulmonary bypass induces leukocyte-platelet adhesion. *Blood* 1992;79:1201–1205.
10. Mangano DT, Siliciano D, Hollenberg M, et al. The Study of Perioperative Ischemia Research Group: Postoperative myocardial ischemia-therapeutic trials using intensive analgesia following surgery. *Anesthesiology* 1992;76: 342–353.
11. Mair J, Weiser C, Seibt I, et al. Troponin T to diagnose myocardial infarction in bypass surgery. *Lancet* 1991;337:434–435.
12. Triggiani M, Dolci A, Donatelli F, et al. Cardiac troponin T and perioperative myocardial damage in coronary surgery. *J Cardiothorac Vasc Anesth* 1995;9:484.
13. Machler H, Metzler H, Sabin K, et al. Preoperative myocardial cell damage in patients with instable angina undergoing coronary artery bypass grafting. *Anesthesiology* 1994;81:1324–1331.
14. Mair J, Larue C, Mair P, et al. Use of cardiac troponin I to diagnose perioperative myocardial infarction in coronary artery bypass grafting. *Clin Chem* 1994;40:2066–2070.
15. Larue C, Calzolari C, Bertinchant JP, et al. Cardiac-specific immunoenzymometric assay of troponin I in the early phase of acute myocardial infarction. *Clin Chem* 1993;39:972–979.

16. Etievent JP, Chocron S, Toubin G, et al. Use of cardiac troponin I as a marker of perioperative myocardial infarction. *Ann Thorac Surg* 1995;59: 1192.
17. Hirsch R, Dent C, Wood MK, et al. Patterns and predictive value of cardiac troponin I after cardiothoracic surgery in children. *Circulation* 1996;94(suppl I):I480.
18. Bodor GS, Porter S, Landt Y, et al. Development of monoclonal antibodies for an assay of cardiac troponin-I and preliminary results in suspected cases of myocardial infarction. *Clin Chem* 1992;38:2203–2214.
19. Immer FF, Stocker F, Seiler A, et al. Troponin I for prediction of early postoperative course after pediatric cardiac surgery. *J Am Coll Cardiol* 1999; 33:1719–1723.
20. Adams JE III, Jeavens AW, Gray LA. Use of cardiac troponin I to detect cardiac injury after cardiac surgery. *Circulation* 1998;99:I93.
21. Greenson NW, Macoviak J, Krishnaswamy P, et al. Troponin I predicts patient outcome after open heart surgery. *Circulation* 1999;100(suppl 1): I291.
22. Hannes W, Seitelberger R, Christoph M, et al. Effect of peri-operative diltiazem on myocardial ischaemia and function in patients receiving mammary artery grafts. *Eur Heart J* 1995;16: 87–93.
23. Krejca M, Skiba J, Szmagala P, et al. Cardiac troponin T release during coronary surgery using intermittent cross-clamp with fibrillation, on-pump, and off-pump beating heart. *Eur J Cardiothorac Surg* 1999;16:337–341.
24. Wendel HP, Heller W, Michel J, et al. Lower cardiac troponin T levels in patients undergoing cardiopulmonary bypass and receiving high-dose aprotinin therapy indicate reduction of perioperative myocardial damage. *J Thorac Cardiovasc Surg* 1995;109:1164–1172.
25. Pelletier LC, Carrier M, Leclerc Y, et al. Intermittent antegrade warm versus cold blood cardioplegia: A prospective, randomized study. *Ann Thorac Surg* 1994;58:41–48.
26. Wan S, Yim APC. Cytokines in myocardial injury: Impact on cardiac surgical approach. *Eur J Cardiothorac Surg* 1999;16(suppl 1):S107-S111.
27. Tate JR, Heathcote D, Rayfield J, Hickman PE. The lack of standardization of cardiac troponin I assay systems. *Clin Chim Acta* 1999;284:141–149.
28. Valdes R, Jortani SA. Standardizing utilization of biomarkers in diagnosis and management of acute cardiac syndromes. *Clin Chim Acta* 1999;284: 135–140.

Chapter 18

The Role of Cardiac Troponin Testing in Renal Disease

Fred S. Apple, PhD

Introduction

Cardiac disease is the major cause of death in patients with end-stage renal disease (ESRD), responsible for 40% to 45% of overall mortality.[1,2] Approximately 20% to 25% of deaths from cardiac causes are due to acute myocardial infarction (AMI). In a recent study by Herzog et al., the overall mortality after AMI among 34,000 patients on long-term dialysis, identified from the US Renal Data System database, was 59% at 1 year, 73% at 2 years, and 89% at 5 years.[1] Furthermore, the mortality rate after AMI was substantially greater for patients on long-term dialysis than for renal transplant recipients. Thus, sudden and cardiac death are common occurrences in hemodialysis patients. Based on approximately 325,000 deaths from 1977 through 1997 identified from the US Renal Data System database, Bleyer et al.[3] demonstrated that for all hemodialysis patients, Monday and Tuesday were the most common days of sudden and cardiac death. Approximately 20% of deaths occurred on Monday and Tuesday, compared with 14% on Wednesday through Saturday. Their conclusions were that the intermittent nature of hemodialysis, accompanied by large weight gains, increased potassium concentrations, and postdialysis hypotension on Mondays and Tuesdays may have contributed to the differences in cardiac death rates. These data support the conclusion that more aggressive strategies are necessary for the prevention and treatment of AMI in patients who are on dialysis.

The purpose of this chapter is to review the literature on the role of monitoring blood cardiac troponin I (cTnI) and T (cTnT) concentra-

From: Adams JE III, Apple FS, Jaffe AS, Wu AHB (eds). *Markers in Cardiology: Current and Future Clinical Applications.* Armonk, NY: Futura Publishing Company, Inc.; © 2001.

tions in ESRD patients for detection of AMI and for use as risk stratification markers for predictors of cardiac events, both during hospitalization and over a 1-year follow-up period. Biochemical and molecular evidence is presented that supports cTnI and cTnT detection in blood originates only from the heart, thus serving as direct evidence of myocardial damage.

The Diagnostic Utility of Cardiac Troponin Testing for Detecting AMI

The diagnostic utility of cTnI and cTnT in renal patients who are on hemodialysis has been challenged. Numerous studies have shown that both markers are increased in serum and plasma in ESRD patients without clinical evidence of myocardial damage. Since 1993, a review of eight studies for cTnI involving a total of 357 ESRD patients[4–11] and nine studies for cTnT involving a total of 534 ESRD patients[4,7–22] showed that 2% to 10% and 10% to 30% had increased troponin values, respectively. Almost all of the several cTnI and the one cTnT diagnostic assay manufacturers show data in their package inserts regarding the percent of ESRD patients that were found to have increased cardiac troponins based on each respective assay. Included are suggested explanations for the cause of these increased troponin concentrations. These explanations include the expression of cTnT in skeletal muscle (which is addressed later in this chapter) as well as the detection of subclinical myocardial damage. However, the cause of the differences in positive rates between cTnT and cTnI is not known. It should be noted that the studies reviewed for cTnT in this chapter involved only the second- or third-generation cTnT assays. It is important to separate out the initial findings involving the first-generation cTnT assay[23] (which has been removed from the marketplace) since this enzyme-linked immunosorbent cTnT assay demonstrated substantial false-positive cTnT results in ESRD patients due to an interference in the assay. Up to 50% to 60% of ESRD patients were shown to have increased cTnT concentrations by the first-generation assay.[4] Assay revisions involving the primary and secondary antibodies decreased blood cTnT increases in ESRD patients to 10% to 30% in the second-generation assay,[24] eliminating any interference by skeletal muscle troponins. Furthermore, a review of the previously mentioned studies re-emphasized the problems of using creatine kinase isoenzyme (CK-MB) as a marker for myocardial damage in ESRD. Falsely increased concentrations in up to 60% to 75% of ESRD patients have been shown,[4,5] most likely due to the re-expression of CK-MB in skeletal muscle[25] because of the myopathy associated with ESRD patients.[26] Thus cTnI and cTnT are markers with high specificity for cardiac damage and should be used to distinguish whether increases in CK-MB are due to myocardial or skeletal muscle injury.[27,28]

Cardiac Troponins as Predictors of Subsequent Cardiac Events

In patients with clinical unstable angina, the presence of increased blood cTnI and cTnT concentrations identifies a subgroup at significantly higher risk of developing cardiac events such as cardiac death and nonfatal AMI, during hospitalization and at long-term follow-ups.[29,30] Several studies have now also demonstrated that ESRD patients with increases in blood cTnI and cTnT concentrations tend to have a poor prognostic outcome.[8,10,12,21,31,32] In a 1-year follow-up study of 16 randomly selected ESRD patients, the cardiac event rate (n = 4 fatal AMIs) was correlated to patients who displayed the higher increases of serum cTnT and cTnI.[10] In another study of 30 ESRD patients, of 10 patients whose serum cTnT exceeded the 0.1 μg/L threshold at entry, 4 were dead at 1 year, and three others were diagnosed with substantial coronary artery disease (CAD).[8] However, only 2 of 30 patients who showed an increased serum cTnI at entry had a cardiac event at 1 year.

Serum cTnT concentrations were also measured in 49 ESRD patients who presented with no complaints of chest pain and in 83 renal insufficiency patients (serum creatinine >2 mg/dL and not on dialysis).[31] All patients were clinically followed for 6 months after entry into the study. Of the 25 ESRD patients with increased cTnT concentrations at entry, 6 had cardiac events. Thus cTnT demonstrated 100% sensitivity and 56% specificity. In comparison, all three patients with an increased cTnI had cardiac events, demonstrating 50% sensitivity and 100% specificity for cTnI. Patients with diabetes were more likely to have increased cardiac troponin concentrations. In contrast, only three patients in the entire renal insufficiency group had an increased cTnI or cTnT. In the 6-month follow-up, two patients suffered an AMI, but neither of these patients had increased troponins. These preliminary data suggest that cardiac troponin testing may be effective in elucidating cardiac risks of patients with ESRD undergoing chronic dialysis.

The measurement of cTnT in the blood of 97 ESRD patients showed that blood cTnT was detectable in 29% of patients.[12] Interestingly, the prevalence of increased cTnT concentrations correlated with cardiac risk. Fifty percent (11 of 22) of known CAD patients had an increased cTnT concentration (median 1.6 μg/L), compared with 31% (15 of 48) in patients with two or more risk factors (median 0.1 μg/L), and 11% (3 of 27) in patients with 0 or 1 risk factor ($P<0.05$ versus known CAD patients). Thus, there appears to be a positive relationship between the increased risk of CAD and increases in cTnT concentrations in blood. Furthermore, a recent study investigated the use of monitoring blood cTnT and cTnI concentrations for predicting cardiac outcomes by 6 months in patients presenting with suspected acute coronary syndromes and renal insufficiency (creatinine >2.0 mg/dL) (n = 51) compared with acute coronary syndrome patients without renal disease (n = 102).[32] Thirty-five percent of patients in the renal group and 45% in the nonrenal group experienced an adverse outcome during initial hospitalization. However, at 6 months, both groups had experienced greater than 50% adverse outcomes. The areas under the receiver operating characteristic curves for both cTnT and cTnI, used as predictors of initial and long-term outcomes, were significantly lower in the renal group than in the nonrenal group (0.56 and 0.75, respectively). Thus, the ability of cardiac troponins to predict cardiac risk for subsequent adverse outcomes in patients presenting with acute coronary syndromes was decreased in the presence of renal disease. Unfortunately, no mechanisms were given to explain these findings. Similar observations were made when serum cTnI monitoring was performed in 84 patients with renal insufficiency hospitalized to rule out AMI.[33] Using clinical parameters (electrocardiography, echocardiography) to diagnose AMI, the clinical sensitivity of cTnI (77%) was significantly better than that of CK-MB (68%). However, the 77% sensitivity observed for cTnI in the renal insufficiency patients[33] was substantially less than the greater than 95% sensitivity observed for cTnI in ischemic chest pain patients without renal insufficiency for ruling in and ruling out AMI.[34] No differences were observed for cTnI regarding clinical specificities: greater than 85% for either renal or nonrenal disease groups.[33,34]

cTnT and cTnI Expression: Heart and Skeletal Muscle

Four isoforms of cTnT have been shown to be expressed in developing cardiac muscle.[35] Cardiac TnT isoforms have also been described in fetal and diseased human skeletal muscle.[4,28,35,36] However, there is a developmental downregulation of cTnT and upregulation of skeletal isoforms of TnT in developing skeletal muscle, which leads to the absence of cTnT in nondiseased adult skeletal muscle. Regarding cTnI, human cardiac muscle contains a single cTnI, and healthy and diseased human fetal and adult skeletal muscle have never been shown to express cTnI.[37]

One reason that was proposed for the unexplained increase of cTnT in the blood of ESRD patients was the possibility of extracardiac expression of cTnT observed in diseased skeletal muscle,[4,35,36] since chronic dialysis is associated with muscle wasting and regenerative processes.[26] In an initial report, cTnT expression in skeletal muscles from ESRD patients was demonstrated (Western blot analysis).[4] However, the antibodies used in these studies were not the same as found in the second- or third-generation cTnT immunoassay kit marketed by Roche Diagnostics (Indianapolis, IN). This report contrasted to a second report by Haller et al.[12] that showed no evidence of the expression of either cTnT messenger ribonucleic acid (mRNA) (reverse-phase polymerase chain reaction) or protein (immunoblot) in truncal skeletal muscle from five ESRD patients. Again, the antibodies used in the immunoblot experiments were not the same as those found in the cTnT Roche assay kit. A third report recently addressed the expression of cTnT isoforms in ESRD skeletal muscles using both the capture antibody (M11.7) and detection antibody (M7) from the Roche cTnT assay kit.[28] The M7 antibody detected a 39 kd cTnT isoform similar to that expressed in human heart tissue in 2 of 45 skeletal muscle biopsies. In contrast, the M11.7 antibody detected 2 to 3 cTnT isoforms at 34 to 36 kd, and no cTnT at 39 kd in 20 of 45 skeletal muscle biopsies. Given the differences in epitopes recognized by the M7 and M11.7 antibodies, it was concluded that the cTnT isoforms expressed in ESRD muscle would not be detected by the Roche cTnT assay if released into the circulation.[28] Therefore, it can be concluded that circulating cTnT and cTnI originate from the heart and are indicative of myocardial damage. Interestingly, though, a very recent study from our laboratory showed expression of a 150 base pair (bp) amplicon in 4 of 7 skeletal muscle biopsies from ESRD patients.[38] This demonstrates, for the first time, cTnT mRNA expression in adult human skeletal muscle corresponding to cTnT isoform expression.

Clinical Implications

The use of cTnT and cTnI testing in blood to assist in ruling in and ruling out AMI and as a tool for cardiac risk assessment in ESRD patients now appears to be evidence based. However, larger trial studies would be useful to determine the overall incidences and differences between cTnT and cTnI monitoring in ESRD patients for cardiac risk assessment. Incorporation of cardiac troponin testing in ESRD patients may assist in initiating more aggressive treatment of CAD, detecting subclinical myocardial injury, and correlating increases in cardiac troponin testing with increased cardiac risk. Moreover, monitoring blood cardiac troponin concentrations should assist in the detection and treatment of CAD before renal transplantation, potentially reducing the risk of adverse cardiac events.

References

1. Herzog CA, Ma JZ, Collins AJ. Poor long-term survival after acute myocardial infarction among patients on long-term dialysis. *N Engl J Med* 1998; 339:799–805.
2. Herzog CA. Diagnosis and treatment of ischemic heart disease in dialysis patients. *Curr Opin Nephrol Hypertens* 1997;6:558–565.
3. Bleyer AJ, Russell GB, Satko SG. Sudden and cardiac death rates in hemodialysis patients. *Kidney Int* 1999;55:1553–1559.
4. McLaurin MD, Apple FS, Voss EM, et al. Cardiac troponin I, cardiac troponin T, and creatine kinase MB in dialysis patients without ischemic heart disease: Evidence of cardiac troponin T expression in skeletal muscle. *Clin Chem* 1997;43:976–982.
5. Adams J, Bodor G, Davila-Roman V, et al. Cardiac troponin I: A marker with high specificity for cardiac injury. *Circulation* 1993;88:101–106.
6. Trinquier S, Flecheux O, Bullenger M, Castex F. Highly specific immunoassay for cardiac troponin I assessed in noninfarct patients with chronic renal failure or severe polytrauma. *Clin Chem* 1995;41:1675–1676.
7. Li D, Jialal I, Keffer J. Greater frequency of increased cardiac troponin T than increased cardiac troponin I in patients with chronic renal failure. *Clin Chem* 1996;42:114–115.
8. Porter GA, Norton T, Bennett WB. Troponin T, a predictor of death in chronic haemodialysis patients. *Eur Heart J* 1998;19(suppl):N34-N37.
9. Bhayana V, Gougoulias T, Cohee S. Discordance between results for serum troponin T and I in renal disease. *Clin Chem* 1995;41:312–317.
10. Apple FS, Sharkey SW, Hoeft P, et al. Prognostic value of serum cardiac troponin I and T in chronic dialysis patients: A one year outcomes analysis. *Am J Kidney Dis* 1997;29:399–403.
11. Hafner G, Thome-Kromer B, Schaube J, et al. Cardiac troponins in serum in chronic renal failure. *Clin Chem* 1994;40:1790–1791.
12. Haller C, Zehelein J, Remppis A, et al. Cardiac troponin T in patients with end-stage renal disease: Absence of expression in truncal skeletal muscle. *Clin Chem* 1998;44:930–938.
13. Ishii J, Ishikawa T, Yukitake J. Clinical specificity of a second generation cardiac troponin T assay in patients with chronic renal failure. *Clin Chim Acta* 1998;270:183–188.
14. Akagi M, Nagake Y, Makino H, Ota Z. A comparative study of myocardial troponin T levels in patients undergoing hemodialysis. *Jpn J Nephrol* 1995; 37:639–643.
15. Frankel WL, Herold DA, Zregler TW, Fitzgerald RL. Cardiac troponin T is elevated in asymptomatic patients with chronic renal failure. *Am J Clin Pathol* 1996;106:118–123.
16. Ooi DS, House AA. Cardiac troponin T in hemodialyzed patients. *Clin Chem* 1998;44:1410–1416.
17. Bhayana V, Gougoulias T, Cohoe S, Henderson AR. Discordance between results for serum troponin T and troponin I in renal disease. *Clin Chem* 1995;41:312–317.
18. Haller C, Stevanovich A, Katus HA. Are cardiac troponins reliable serodiagnostic markers of cardiac ischaemia in end-stage renal disease? *Nephrol Dial Transplant* 1996;11:941–944.
19. Katus HA, Haller C, Müller-Bardorff M, et al. Cardiac troponin T in end-stage renal disease patients undergoing chronic maintenance hemodialysis. *Clin Chem* 1995;41:1201–1202.

20. Haller C, Katus HA. Cardiac troponin T in dialysis patients. *Clin Chem* 1998;44:358.
21. McNeil AR, Marshall M, Ellis CJ, Hawkins RC. Why is troponin T increased in the serum of patients with end-stage renal disease. *Clin Chem* 1998;44: 2377–2378.
22. Collinson PO, Stubbs PJ, Rosalki SB. Cardiac troponin T in renal disease. *Clin Chem* 1995;41:1671–1673.
23. Katus HA, Looser S, Hallermayer K, et al. Development and in vitro characterization of a new immunoassay of cardiac troponin T. *Clin Chem* 1992; 38:386–393.
24. Müller-Bardorff M, Hallermayer K, Schröder A, et al. Improved troponin T ELISA specific for cardiac troponin T isoform: Assay development and analytical and clinical validation. *Clin Chem* 1997;43:458–466.
25. Apple FS, Billadello JJ. Expression of creatine kinase M and B mRNAs in treadmill trained rat skeletal muscle. *Life Sci* 1994;55:585–592.
26. Diesel W, Emms M, Knight B, et al. Morphology features on the myopathy associated with chronic renal failure. *Am J Kidney Dis* 1993;22:677–684.
27. Bodor GS, Porterfield D, Voss EM, et al. Cardiac troponin I is not expressed in fetal and healthy or diseased adult human skeletal tissue. *Clin Chem* 1995;41:1710–1715.
28. Ricchiuti V, Voss EM, Ney A, et al. Cardiac troponin T isoforms expressed in renal diseased skeletal muscle will not cause false-positive results by the second generation cardiac troponin T assay by Boehringer Mannheim. *Clin Chem* 1998;44:1919–1924.
29. Ohman EM, Armstrong PW, Christenson RH, et al. Cardiac troponin T levels for risk stratification in acute myocardial ischemia. *N Engl J Med* 1996;335:1333–1341.
30. Galvani M, Ottani F, Ferrini D, et al. Prognostic influence of elevated values of cardiac troponin I in patients with unstable angina. *Circulation* 1997;95: 2053–2059.
31. Roppolo LP, Fitzgerald R, Dillow J, et al. A comparison of troponin T and troponin I as predictors of cardiac events in patients undergoing chronic dialysis at a Veteran's hospital: A pilot study. *J Am Coll Cardiol* 1999;34: 448–454.
32. VanLente F, McErlean ES, DeLuca SA, et al. Ability of troponins to predict adverse outcomes in patients with renal insufficiency and suspected acute coronary syndromes: A case matched study. *J Am Coll Cardiol* 1999,33: 471–478.
33. McLaurin MD, Apple FS, Falahati A, et al. Cardiac troponin I and creatine kinase MB mass to rule out myocardial injury in hospitalized patients with renal insufficiency. *Am J Cardiol* 1998;82:973–975.
34. Falahati A, Sharkey SW, Christenson D, et al. Implementation of cardiac troponin I for detection of acute myocardial injury in an urban medical center. *Am Heart J* 1999;137:332–337.
35. Anderson PAW, Malouf NN, Oakley AE, et al. Troponin T isoform expression in humans: A comparison among normal and failing adult heart, fetal heart, and adult and fetal skeletal muscle. *Circ Res* 1991;69:1226–1233.
36. Bodor GS, Survant L, Voss EM, et al. Cardiac troponin T composition in normal and regenerating skeletal muscle. *Clin Chem* 1997;43:476–484.
37. Wilkinson JM, Grand JA. Comparison of amino acid sequence of troponin I from different striated muscles. *Nature* 1978;271:31–35.
38. Ricchiuti V, Apple FS. RNA expression of cardiac troponin T isoforms in diseased human skeletal muscle. *Clin Chem* 1999;45:2129–2135.

Chapter 19

The Use of Cardiac Biomarkers for the Detection of Drug-Induced Myocardial Damage

Eugene H. Herman, PhD and V.J. Ferrans, MD, PhD

With the development of numerous new pharmaceuticals and the widespread abuse of certain types of drugs, the potential for toxic myocardial reactions can be a serious problem. The most definitive technique for documenting drug-induced cardiac disease is the endomyocardial biopsy.[1] However, routine use of this invasive procedure is not practical, and thus there is a continuing need for reliable noninvasive methods to detect drug-induced myocardial damage. Most of the methods currently in use for this purpose are based on the presence in serum of cellular proteins and other components that are released from damaged myocytes or activated endothelial cells into the circulating blood. These substances can be classified into two categories, according to whether they are normally present in small concentrations in serum. For specificity, it is necessary to determine, by appropriate electrophoretic or immunologic assays, whether the increased serum levels of these components are a consequence of their release from cardiac myocytes, skeletal muscle cells, or other types of cells. If exclusively myocardial, quantification of the amount released provides a measure of the extent of the cardiac damage. The natriuretic peptides represent an exception to this generalization since increases in their plasma levels constitute a response to cardiac dysfunction rather than a direct indication of myocyte damage.

Aspartate Aminotransferase and Lactic Dehydrogenase

Biochemical assays for the detection of myocardial injury were first reported by Karmen et al.[2] and La due et al.[3] These investigators

From: Adams JE III, Apple FS, Jaffe AS, Wu AHB (eds). *Markers in Cardiology: Current and Future Clinical Applications*. Armonk, NY: Futura Publishing Company, Inc.; © 2001.

observed that the serum concentration of aspartate aminotransferase (AST; previously referred to as glutamic-oxaloacetic transaminase or SGOT), an enzyme normally present in the cytosol of the myocyte, was elevated in patients with acute myocardial infarction (AMI). Shortly thereafter, the serum levels of lactic dehydrogenase (LDH) were also found to be increased after AMI.[4] The reliability of this assay was improved by electrophoretic study of the LDH isoenzymes which showed LDH1 to be the predominant form found in serum after AMI.[5]

Creatine Kinase

Creatine kinase (CK) was the third myocyte cytosolic enzyme found to show increased serum activity after AMI.[6] The sensitivity of this assay was also improved by monitoring the MB isoenzyme of CK or CK-MB mass.[7–9] Cardiac markers such as CK and CK-MB have been used in clinical studies to detect myocardial damage induced by various drugs and toxic agents. Much of the early information concerning the diagnostic value of CK or other enzymes in assessing these types of damage has been largely derived from isolated case studies. The usefulness of these enzymes as biomarkers for the evaluation of drug-induced myocardial damage is limited because the amount of acute cardiac muscle injury produced by most drugs is much smaller than that which occurs in AMI. Furthermore, the timing of the release of these components also may differ markedly in these two conditions, thus adding to the difficulty of evaluating data of this type.

The serum enzymes, CK-MB in particular, have had their most consistent application as biomarkers for the detection of AMI. The specificity of these measurements is limited because significant amounts of CK-MB are present in skeletal muscle.[10] In addition, drugs such as benzodiazepines, tricyclic antidepressants, pyridoxine, and high doses of acetylsalicylic acid can cause false elevations of CK-MB.[11] Because of problems in the interpretation of CK-MB data, interest became focused on the use of the myocyte myofibrillar proteins, troponin T and troponin I, as markers of myocardial injury.[12,13]

Troponins

The three troponins, namely, troponin C (18 kd), troponin T (37 kd), and troponin I (26 kd), are structural proteins that are bound to the actin (thin) filaments and regulate the calcium-mediated interaction of actin and myosin in striated muscle.[14] All three of these proteins are products of different and unrelated genes. Troponin T and troponin I exist in three different isoforms (which are localized in cardiac, fast-twitch, and slow-twitch muscle).[15,16] Cardiac troponin I (cTnI) contains 31 amino acid residues at the N-terminus that are not found in either the fast or slow skeletal muscle isoforms. There is approximately a 40% dissimilarity between the structure of cTnI and that of the other two isoforms.[15] cTnI appears to be the only troponin isoform expressed in the myocardium, even in the presence of chronic disease processes.[17] These properties are considered favorable for a specific biomarker of myocardial injury.

In contrast, cardiac troponin T (cTnT) differs by only 6 to 11 amino acid residues from the skeletal muscle isoforms.[16] Both cardiac and skeletal TnT are coexpressed in the fetal heart, with the skeletal muscle isoform being suppressed during ontogeny.[18] The skeletal muscle form can be re-expressed in the human myocardium under conditions of cardiac stress.[18] The second-generation cTnT assay (enzymun troponin T) uses an anti-cTnT antibody that is specific for the cardiac isoform and does not cross-react with skeletal muscle TnT.[19] Cardiac TnT is also expressed in fetal skeletal muscle and has been shown to be re-expressed in regenerating rat skeletal muscle after injury or denervation.[20] There is some evidence to indicate that cTnT isoforms can be re-expressed in the skeletal muscle of patients who have certain diseases that affect the skeletal muscles or have end-stage renal disease.[21] The potential reappearance of cTnT isoforms has raised concerns about the usefulness of assays for this cardiac biomarker in patients with such diseases. However, Ricchiuti et al.[22] have shown that the cTnT isoforms re-expressed in patients with renal or skeletal muscle diseases will not cause falsely elevated values when cTnT determinations are made with the second-generation commercial assay. The improved third-generation cTnT assay retains both specificity and sensitivity but requires less time to perform.[23]

The troponins are found in at least two intracellular pools. Small amounts of cTnT and cTnI (3% to 7%) are present in the cytoplasm, and the remainder is complexed with actin.[24,25] The small cytoplasmic pool is released rapidly after myocardial injury and is associated with early diagnostic sensitivity, whereas the remaining large store of actin-bound troponin provides the basis for longer term diagnostic studies. In contrast to CK and other enzyme markers, the baseline levels of cTnT and cTnI normally are extremely low (below or at the limit of sensitivity of the assay), and even small elevations in the blood level of these markers represent definitive evidence of myocardial injury.[21] The increased specificity and sensitivity of cTnT and cTnI assays have expanded their clinical utility from diagnosing AMI to detecting cardiac injury after thrombolytic therapy, cardiac and noncardiac operations, cardiac reperfusion, heart transplantation, and heart contusions.[25]

The use of cardiac troponins has also been advocated for detecting subtler myocardial injury related to myocarditis and exposure to cardiotoxic agents, and for determining the efficacy of cardioprotective procedures. Interest in diagnostic biomarkers for the cardiac toxicity of new drugs arises from the need to identify these myocardial alterations in the preclinical development of these agents. Holt[26] has provided a synopsis of current concerns regarding the nonclinical use of biomarkers and has identified potential problems that might limit their use in diagnosing myocardial injury in animals. Some of these concerns have already begun to be addressed. The clinical immunoassays developed for humans to detect the cardiac forms of troponin T and troponin I can be used to analyze blood samples collected from animals. The cardiac troponins are found in high concentration in the myocardium of vertebrates. Epitopes of these proteins are conserved in the troponin structure across a number of animal phyla.[27]

The question of the cross-reactivity between human and animal troponins in bioassays was examined by O'Brien et al.,[28] who determined that the antibodies used in the clinical first- and second-generation cTnT immunoassays also recognized epitopes found in animals. They found that antibodies raised against human cTnT also showed reactivity against cTnT from mammals, birds, and fish. Rat hearts had the highest reactivity, and chicken and fish hearts had the lowest reactivity. Skeletal muscle from rats, pigs, and goats had 10% of the reactivity of cTnT with the first-generation assay and 1% of the cTnT reactivity with the second-generation assay.[28] They concluded that in animals the second-generation assay had sufficient reactivity and selectivity to differentiate between myocardial and skeletal muscle damage.[28] Only one cTnT assay (which is standardized) is now commercially available, in contrast to several cTnI assays (with no standardization consensus). At present, it is not clear whether the antibodies used in the latter assays, which recognize different epitopes of the human cTnI molecule, would be useful for the detection of cTnI in animals.

Assays for cTnT and cTnI have been used to detect myocardial damage due to ischemia in rats and dogs,[29–31] immunologic injury in rats,[32] myocarditis in mice,[33,34] and drug toxicity in rats and mice.[31,35] Cardiac TnT was monitored in all studies except in that of Smith et al.[34] which used the cTnI assay. The findings in these studies support the concept that monitoring serum cTnT levels can be used as a sensitive means to detect myocardial injury induced in laboratory animals.

Natriuretic Peptides

Under normal conditions, both atrial natriuretic peptide (ANP) and brain natriuretic peptide (BNP) are detected in the plasma.[36,37]

Increased concentrations of ANP and BNP have been detected in patients with various types of heart disease.[38] Monitoring the levels of these peptides is useful in evaluating the severity of congestive heart failure.[38,39] Both ANP and BNP exert natriuretic, diuretic, and vasodilating actions and are released in response to increases in systemic blood pressure and plasma volume.[40] They are therefore indicators of cardiac dysfunction, not of cardiac damage. Upon secretion from the atria, the 126 amino acid ANP prohormone is activated into the biologically active ANP and the N-terminal fragment (N-terminal pro-ANP) (NT pro-ANP).[41] These two peptides are secreted in equimolar amounts, but owing to its longer half-life, the NT pro-ANP is found in greater concentrations than is ANP.[41] Human BNP is synthesized in the brain and the ventricle of the heart as a 108 amino acid precursor (pro-BNP). A low molecular weight form of BNP (BNP 32) is the biologically active portion of the peptide and is the major circulating form of BNP.[42] Detection of ANP and NT pro-ANP requires extraction of the peptides from the plasma before analysis by radioimmunoassay.[38] Due to methodological differences in the extraction procedure and in antibody specificity toward the peptides, baseline concentrations of ANP in normal subjects have varied as much as six-fold. Both radioimmunoassay and immunoradiometric assay to monitor the plasma concentration of BNP are commercially available.[43] However, because of differences in methodologies, antibodies, and tracers, no international BNP standard has been established.

Anthracyclines

Various manifestations of myocardial toxicity (ischemia, infarction, arrhythmias, myocarditis, pericarditis, and cardiomyopathy) have been observed after treatment with certain types of antineoplastic agents. In some instances these effects become apparent during or shortly after dosing (5-fluorouracil), whereas in others the toxicity is not apparent until some time has elapsed after the completion of therapy (anthracyclines). The use of a biomarker that is capable of detecting this toxicity could have a significant effect on the ultimate course of therapy with either type of agent.

Anthracyclines such as doxorubicin (DXR) are among the most effective chemotherapeutic agents, but their optimal clinical use is limited by a dose-dependent cardiomyopathy.[44] A number of noninvasive methods (echocardiography, radionuclide ventriculography, electrocardiography [ECG]) have been used to detect and evaluate the extent of DXR-induced myocardial damage.[45] These techniques have not been able to detect the toxicity as it begins to evolve.[45] There is a need for other noninvasive methods that can detect initial stages of the cardiotoxicity to allow for adjustments in the dose of the chemotherapy. Attempts to monitor DXR-induced cardiomyopathy have included the measurement of serum levels of CK and other substances as cardiac biomarkers.

Anthracyclines have been found to alter the levels of CK in rodents. Acute high doses of DXR caused significant elevations in total serum LDH, LDH1, LDH2 and LDH3, and the LDH1:LDH2 ratio.[46] Because of widespread tissue injury, these findings were thought to be too nonspecific to indicate a cardiac origin for the elevation of these enzyme levels. Other studies have shown that single, high doses of DXR produced significant increases in the serum levels of CK and other enzymes in rodents. Many investigators have attempted to use changes in CK activity after treatment with DXR in rodents as one means of identifying potential cardioprotective agents. For example, Bhanumathi et al.[47] and Dobric et al.[48] reported increased CK activity after the administration of an acute high dose of DXR to mice and rats. This increase was significantly attenuated by pretreatment with the sulfhydryl compounds WR1605 or WR2729. However, morphologic confirmation of cardioprotection was not provided. De Leonardis et al.[49] found no significant change in the serum levels of CK-MB in seven patients who had received cumulative DXR doses between 200 and 490 mg/m^2. This lack of change in serum CK-MB levels was attributed to the coadministration of L-carnitine. Since there was no comparable group of patients receiving DXR alone, it is not known whether the results indicate a cardioprotective effect of L-carnitine or a lack of sensitivity of the CK-MB assay. No change in serum CK-MB levels was reported in other clinical trials (in children and adults) after treatment with various cumulative doses of DXR.[50,51]

The application of cTnT as a biomarker of DXR cardiotoxicity was first reported in spontaneously hypertensive rats (SHR). Seino et al.[52] detected elevated serum cTnT levels in SHR that were given 8–weekly doses of 1.5 mg/kg DXR. A subsequent clinical study found no change in cTnT levels in children who had received three to five doses of DXR, daunorubicin, or idarubicin chemotherapy.[50] These investigators may not have been able to detect small increases of cTnT because the criterion for elevation was set at a level of 0.2 μg/L or greater. Raderer et al.[53] monitored serum cTnT levels for up to 48 hours in patients after initial doses of DXR (50 mg/m^2) or epirubicin (100 mg/m^2). Again no rise in cTnT levels was detected. They interpreted these results as indicating that anthracycline-induced cardiotoxicity does not evolve from acute myocyte damage. In contrast, Ottinger et al.[54] found that serum levels of cTnT increased from nonmeasurable to low levels in children receiving DXR. Using the second-generation assay, Lipshultz et al.[55] observed that the low-level increases in serum concentrations of cTnT (0.03 to 0.09 μg/L) detected after the initial induction dose of DXR or subsequent intensification doses (45 to 222 mg/m^2) were predictive of risk for left ventricular abnormalities, including dilatation and wall thinning in children. Increases in serum levels of cTnI were reported by Missov et al.[51] Mean cTnI values of 71.3 ± 29.2 pg/mL were detected in patients treated with intermediate cumulative doses of DXR (240 to 300 mg/m^2) compared with 35.8 ± 17.5 pg/mL in patients with cancer not receiving anthracyclines. For undetermined reasons, cTnI levels were elevated (17.5 ± 17.9 pg/mL) in some of their normal control subjects. Studies by Herman et al.[56,57] have confirmed the utility of monitoring serum cTnT to detect the extent of DXR-induced cardiac toxicity. These investigators found that the magnitude of the increase in serum levels of cTnT correlated with the total cumulative dose of DXR and with the severity of the cardiac lesion scores as determined by microscopic examination of myocardial tissue. The antibody used in the cTnT assay does specifically stain myofibrillar-associated cTnT in myocytes. The intensity of this staining was decreased in the myocytes of SHR that were treated with DXR and had increased serum levels of cTnT.[56,57] Adamcova et al.[58] evaluated the diagnostic utility of cTnT as a biomarker of myocardial damage caused by daunorubicin in rabbits. They administered 3 mg/kg daunorubicin/week for 8 weeks and found that cTnT levels were normal (defined as <0.1 μg/L) through the fifth dosing cycle. The initial pretreatment values were nearly zero, and small but significant increases in serum cTnT could be found before the fifth dosing. Mean cTnT levels of 0.22 ± 0.08 μg/L were detected in three of seven animals that had severe myocardial lesions and died after the eighth dose.

The amounts of cTnT released into the circulation appear to vary according to the type of myocyte injury. After AMI, the serum cTnT concentrations reach levels of several micrograms per liter.[25] Rats and dogs subjected to ischemia/reperfusion had elevated cTnT levels.[31] After 90 minutes of coronary occlusion, the serum concentrations of cTnT increased to 13 μg/L after 4.5 hours of reperfusion in dogs and 100 μg/L after 130 minutes of reperfusion in rats.[31] The serum cTnT concentrations reached levels of 3.75 μg/L within 6 hours in rats developing myocardial necrosis after two doses of isoproterenol.[35] Considerably less cTnT is released from myocytes damaged by anthracyclines than from those with isoproterenol-induced necrosis. The highest cTnT concentration reported by Herman et al.[57] was 0.66 μg/L in an animal that received a total of 12 mg/kg DXR and had a maximal cardiomyopathy score of 3. Cardiac TnT levels of 0.30 μg/L or less were observed in all other SHR with DXR-induced myocardial lesions.[57] The final sampling from daunorubicin-treated rabbits detected a mean serum cTnT concentration of 0.13 μg/mL.[58] Cardiac TnT serum levels increased from 0.01 μg/L to 0.03 to 0.09 μg/L in children treated with DXR. These data were clinically meaningful as predictors of cardiotoxicity.[54,55] Thus, small changes in cTnT after anthracycline therapy provide useful diagnostic information.

Considerable interest has developed as to whether changes in plasma concentrations of ANP or BNP, or both, are useful in assessing cardiac function during the course of anthracycline chemotherapy. Bauch et al.[59] were the first to report a change in ANP in association with DXR therapy. They monitored plasma levels of ANP in 16 patients (5 to 19 years of age) and found that significant increases occurred in six patients (that is, 136.2 ± 23.3 pg/mL versus 33.3 ± 4.1 pg/mL in the other 10 patients) within 3 weeks after the last dosing. Five of the six patients had received between 160 and 370 mg/m^2 DXR, and two developed congestive heart failure. Tikanoja et al.[60] monitored serum levels of N-terminal ANP (NT-ANP) in children during treatment (43 patients) and after treatment (48 patients) with DXR. During treatment, the serum NT-ANP concentration increased over those in age-matched controls (0.26 versus 0.14 nmol/L), but did not correlate with the cumulative doses of DXR. Blood collection times were not standardized and, as a result, NT-ANP levels could be affected by a diurnal variation.[61] Tikanoja et al.[60] concluded that monitoring serum NT-ANP levels during chemotherapy was of relatively minor diagnostic utility. In a second group of patients who had previously completed chemotherapy for a median of 5 years, these investigators found serum levels of NT-ANP to be higher than those in age-matched controls (0.22 versus 0.14 nmol/L). The highest levels (0.30 nmol/L) were found in patients who received bone marrow transplantation or cardiac irradiation, or both. Monitoring the serum concentration of NT-ANP might be helpful in the overall long-term assessment of cardiac function in patients who have completed chemotherapy. Toyoda et al.[62] used 5-weekly intracoronary infusions of DXR to produce an experimental model of dilated cardiomyopathy. Three months after the final infusion, significant changes in cardiac structure and function were noted. During this same period, plasma ANP levels increased from 33.8 ± 7.0 to 76.5 ± 14.8 pg/mL. Yokota et al.[63] compared the plasma levels of BNP in normal rats and in rats that developed a nephrotic syndrome after the administration of a single high dose of DXR (7 mg/kg). Plasma levels of BNP rose with time and more than doubled by 3 weeks after dosing (2.3 ± 0.6 versus 0.8 ± 0.2 fmol/mL). Parallel increases were also noted in the plasma ANP concentrations. An examination of cardiac structure and function was not included in this study. Suzuki et al.[64] examined the potential diagnostic role of BNP in 27 adult patients who had received an average cumulative dose of 221 ± 54 mg/m^2 of DXR. Except for three patients, transient BNP increases reached maximal levels within 3 to 7 days after treatment and then returned to baseline levels over a 2-week period. Two of the three patients with persistently elevated BNP levels eventually died from cardiac failure. It should be noted that circulating levels of other hormones (ANP, renin, aldosterone, norepinephrine, and epinephrine) and myocardial markers (CK-MB and myosin light chain) did not become abnormal. The plasma levels of ANP, NT pro-ANP, and BNP were monitored in 28 patients undergoing DXR-containing therapy.[65] Clinical evidence of heart failure and increases in the natriuretic peptides were observed in two patients. A significant correlation between increased plasma ANP and decreased left ventricular ejection fraction (LVEF) and a trend toward significant correlation between the increased NT pro-ANP and decreased LVEF were observed in these two patients. Decreases in LVEF began early in therapy at cumulative doses of 200 mg/m^2 DXR, whereas the increases in plasma natriuretic peptides did not become apparent until a 400 mg/m^2 cumulative dose of DXR had been reached. It appears that in this study, serial measurements of natriuretic factors were not useful for predicting left ventricular dysfunction but could be used to detect ongoing subclinical left ventricular dysfunction in patients receiving DXR therapy. In a retrospective study, Yamashita et al.[66] compared ANP and endothelin 1 (ET-1) levels as potential biomarkers for DXR cardiotoxicity. Plasma levels of ET-1 are increased in patients with congestive heart failure.[67] The plasma concentrations of ET-1 rose progressively in 5 of 30 patients during DXR therapy (8.0 versus 2.5 pg/mL).[66] Two of these patients developed congestive heart failure after receiving cumulative DXR doses of 480 to 500 mg/m^2. Serial measurement of plasma ANP and measures of cardiac function (fractional shortening and LVEFs) remained within normal limits until the patients developed overt congestive heart failure. Another group of 25 patients treated with 400 to 600 mg/m^2 without developing congestive heart failure were found to have normal levels of ANP and ET-1 and normal cardiac function.[66] These investigators suggest that ET-1 is a more reliable marker than ANP, but additional studies are still needed to better define whether the natriuretic peptides or ET-1, or both, could be used as reliable indicators of myocardial status after anthracycline therapy.

5-Fluorouracil

5-Fluorouracil (5-FU) is an antineoplastic agent that infrequently causes angina or MI, or both.[68] The risk for these adverse reactions tends to be greater in patients with a history of ischemic heart disease. Experience with a cardiac biomarker to detect the toxicity has been very limited and has involved monitoring serum CK levels. In one study, a patient experienced chest pain during each of two 5-day infusions of 5-FU.[69] During the periods of pain, the ECG and CK activity remained normal. At autopsy, shortly after the treatment cycles had been completed, two recent MIs were detected in this patient. Another patient had normal CK levels before the start of 5-FU therapy.[70] The serum CK levels began to rise after 4 weeks of treatment. In this instance, isoform analysis showed that the increased amounts of CK originated from the skeletal muscle. A 4.2% incidence of 5-FU-induced cardiotoxicity was detected in a prospective study involving 104 patients.[71] As was found in the two other reports cited above, monitoring serum cardiac enzymes such as CK was not helpful in identifying patients experiencing myocardial alterations due to infusion of 5-FU.

Sympathomimetic Agents

Phenylpropanolamine and pseudoephedrine are used as decongestants. Phenylpropanolamine is also used as an appetite suppressant. The pharmacologic effects of these agents are mainly due to direct stimulation of α-adrenergic receptors, and/or indirectly through stimulation of both α- and β-adrenergic receptors by release of endogenous catecholamines.[72,73] Pentel et al.[74] described three patients with phenylpropanolamine intoxication who presented with increased serum CK-MB activity and ECG abnormalities. These findings were interpreted as indicative of myocardial injury. Subsequently, Leo et al.[75] reported two additional cases of suspected cardiac injury (elevated CK and CK-MB levels and abnormal ECG) patterns, which they attributed to the use of anorectic products containing phenylpropanolamine. Neither of these patients had a previous history of cardiovascular disease. Wiener et al.[76] reported that a young man with no known cardiac risk factors developed chest pain and acute ST-segment elevation after the ingestion of pseudoephedrine. CK and CK-MB fraction increased to levels that were thought to indicate myocardial injury. It was suggested that pseudoephedrine induced myocyte injury by α-receptor stimulation in the coronary arteries.

Administration of high doses of the nonspecific β-adrenergic agonist isoproterenol has been used to induce myocardial necrosis in rats. Biomarkers such as total CK and LDH have been used in this experimental model to assess the degree of myocardial damage. Some investigators have detected increases in total CK after treating the animals with very large doses of isoproterenol (100 to 500 mg/kg)[77,78] whereas in other instances, a 5 mg/kg dose of isoproterenol was sufficient to cause necrosis and elevations in both total CK and LDH.[79] Preus et al.[80] monitored CK and LDH and their isoenzymes α-hydroxy butyrate dehydrogenase (α-HBDH) and the LDH:α-HBDH ratio after administering single doses of 5 to 250 mg/kg isoproterenol to Wistar rats. Significant increases in serum levels of CK-MB, LDH1, LDH2, and α-HBDH, and a decrease in the LDH:α-HBDH ratio were detected 2 hours after isoproterenol and reached maximum levels by 4 to 8 hours after dosing. In most instances, the increases in enzyme activity were not dose-dependent. Myocardial necrosis was confirmed by histologic study. These investigators concluded that monitoring the serum activities of CK-MB alone or LDH1–2 isoenzymes in combination with other tests to rule out kidney damage (because of high intrinsic levels of renal LDH1 and LDH2 activities) was the best way to evaluate isoproterenol-induced myocardial lesions in rats.

In contrast, Bleuel et al.[35] did not detect significant changes in CK and LDH levels after the administration of two doses of 4 mg/kg isoproterenol to 7-week-old Sprague-Dawley rats. However, these investigators did detect significant increases in cTnT (up to a mean of 3.75 μg/L versus 0.1 μg/L in control animals) within 6 hours. These levels were still above control values 48 hours after dosing. Marked myocardial damage (evaluated by light microscopy) was detected in all animals in which serum concentrations of cTnT exceeded 2 μg/L. Bertsch et al.[81] compared the effect of the β-receptor agonist orciprenaline (5 mg/kg) on serum levels of cTnT and cTnI. Increases in cTnT levels of 7.8 times and in cTnI of 13.2 times over control values were found 6 hours after dosing. Both cTnT (7.2×) and cTnI (6.0×) remained elevated for up to 24 hours. They concluded that cTnI, like cTnT, can be used as a cardiac marker and that an increase to at least 4.1 μg/L is necessary to detect the marked cardiac damage in this rat model. The cTnT and cTnI immunoassays were used to monitor levels of the biomarkers in this study. Bertsch et al.,[82] using this same animal model, found that cardiac damage could be detected not only by immunoassay but also by a recently developed commercial cTnT rapid test strip. Within 1 to 2 hours after treatment, increases in serum cTnT could be detected by both methods. A positive rapid test strip response occurred when cTnT concentrations reached 0.64 μg/L or higher. The investigators concluded that the rapid test strip is most useful in the initial phase of the myocardial injury (when cTnT concentrations are higher) and that it could give false-negative results when too much time has elapsed since the onset of cardiac damage.

Stimulants of β-adrenergic receptors are used as tocolytic agents to prevent premature birth. These agents are not completely selective for the smooth muscle β-2 receptors and thus could also cause cardiac overstimulation and cardiotoxicity through activation of adrenergic β-1 receptors.[83] This possibility prompted the evaluation of CK and CK isoenzyme levels in pregnant women undergoing tocolytic treatment. Initial reports indicated that the prevention of premature labor with fenoterol for periods of up to 14 weeks did not cause increases in CK-MB levels.[84–86] Newborn children whose mothers had received long-term fenoterol treatment also were found to have normal levels of CK-MB and no evidence of myocardial damage.[87] The levels of CK-MB and LDH isoenzymes remained within the normal range in pregnant women treated with ritodrine, even though these women developed ECG changes during treatment.[88] Previously, Gerris et al.[89] had reported significant increases in the serum activity of the MB, BB, and MM CK isoenzymes during the postpartum period after tocolysis with ritodrine. In most instances, no difference was noted between the isoenzyme distribution pattern associated with AMI and with normal labor. The monitoring of potential cardiotoxicity induced by tocolytic agents is problematic since both CK and CK-MB are found in the uterus and placenta.[90] Levels of these enzymes are increased in women without tocolytic treatment on the first postpartum day.[90]

Adamcova et al.[91] first reported the analysis of cTnT in the cord blood of 15 neonates exposed in utero to terbutaline (Bricanyl; Hoechst Marion Roussel, Kansas City, MO) for preventive tocolysis. Serum levels greater than 0.1 μg/L were detected in four of the neonates. These investigators found no correlation between the levels of cTnT and the levels of CK or CK-MB. Adamcova et al.[92] also used the cTnT immunoassay to examine the potential cardiotoxic effects of infusions of the β-agonist fenoterol and of verapamil. They found a slight increase in cTnT levels during the first day of treatment (0.1 ± 0.03 μg/L versus 0.08 ± 0.01 μg/L in healthy 32- to 36-week pregnant women) and a further increase to 0.35 ± 0.14 μg/L by the third day of tocolytic therapy. They were also able to monitor cTnT levels in the cord blood of neonates born on the third day of tocolytic therapy. In this instance, the concentrations in the cord did not correspond with those found in the mother. The relationship between the elevated cTnT levels and actual ongoing myocardial injury was not addressed in this study.

β-Adrenergic agonists are also used to control the symptoms of asthma. CK was monitored in 13 patients with recurrent asthma who were being treated with subcutaneous terbutaline and in a comparable group of patients with moderate asthma treated with inhalations of terbutaline.[93] The median serum total CK activity was higher in the group receiving subcutaneous terbutaline. The CK-MB level was not increased in either group, and the source of the raised CK concentrations in the group receiving terbutaline subcutaneously was thought to be of noncardiac origin. CK-MB concentrations were determined in 15 children who had received intravenous terbutaline, with or without epinephrine, as emergency treatment for status asthmaticus.[94] No relationship was found between the levels of CK-MB and the doses of either terbutaline or epinephrine.

Theophylline

Animals given large doses of theophylline develop vascular and hemorrhagic lesions, as well as focal myocardial necrosis (particularly in the atria).[95–97] The sera from 30 patients who received high doses of theophylline were analyzed for total CK, CK-MB, and cathodally migrating CK.[98] The cathodally migrating CK form was detected in 11 patients, none of whom had CK-MB levels exceeding the threshold considered indicative of MI. A patient with severe theophylline intoxication had increased serum levels of both CK and CK-MB.[99] In this instance, the pattern of elevation in CK-MB activity was similar to that occurring after AMI. However, the clinical findings in this patient were not consistent with the occurrence of myocardial damage. The effect of intravenous theophylline on CK and CK-MB activity and the frequency of myocardial arrhythmias was assessed in 12 patients with bronchial asthma and spastic bronchiolitis.[100,101] Tachycardia was a common finding. The CK-MB level was elevated 5 minutes after the infusion of theophylline was terminated, and remained elevated for up to 3 days. Clinically silent ventricular arrhythmias developed in four patients who had high CK-MB and theophylline levels, but no overt myocardial damage was reported.

Cocaine

Cocaine abuse is known to induce focal myocardial necrosis and fibrosis.[102] However, the correlation of these findings with chest pain and other signs and symptoms of heart disease is difficult. Cocaine is a widely abused drug which can provoke alterations in the cardiovascular and other organ systems.[103] It can cause chest pain and other symptoms of ischemic heart disease as well as potentially fatal arrhythmias.[104] The initial clinical and ECG evaluation of cocaine users experiencing chest pain is aimed at excluding MI.[105,106] The use of cardiac biomarkers has been advocated to increase the likelihood of detecting cocaine-induced myocardial injury.

Tokarski et al.[107] examined 42 patients with cocaine-associated chest pain. They detected initial CK-MB elevations in 19% of the patients, but only 5% exhibited a CK-MB pattern, which has been reported to occur in conjunction with AMI. Amin et al.[108] identified AMI (22 patients) and transient myocardial ischemia (9 patients) in a group of

50 patients hospitalized with chest pain after cocaine use. CK levels were increased in 75% of the patients, including a significant number that did not develop AMI. They found CK-MB to be predictive, as elevations in this isoenzyme also occurred in patients with AMI. Gitter et al.[106] detected elevations in CK and in total CK but not CK-MB in more than 40% of 101 patients experiencing cocaine-related chest pain. None of these patients were ultimately diagnosed as having AMI. Hollander et al.[105] detected increased levels of CK-MB in approximately 50% of 346 patients with cocaine-induced chest pain. However, they could confirm MI in only 14 of these patients.[105] Rubin and Neugarten[109] found increased CK levels in seven patients who were hospitalized with chest pain after the use of cocaine. A cardiac origin for the observed elevation in CK activity was excluded because CK-MB levels remained normal. The chest pain in these patients was attributed to rhabdomyolysis of thoracic skeletal muscle, a toxicity described previously.[110] McLaurin et al.[111,112] also have suggested that this action, rather than cardiac effects, may be responsible for the elevated levels of CK and CK-MB they observed in patients with cocaine-related chest pain. The absence of myocardial injury was confirmed by the finding that levels of cTnI and cTnT were within the normal range in these patients. Hollander et al.[113] determined that recent cocaine use affected the specificity of myoglobin more than that of CK-MB (but did not alter the sensitivity of cTnI) for the detection of the subsequent occurrence of AMI. They concluded that cTnI appears to be the most specific biomarker available for the detection of myocardial injury induced by the recent use of cocaine.

Other Drugs that Affect the Central Nervous System

Assaying various serum enzyme activities appears to be of limited value in many instances of drug overdose. Wright et al.[114] monitored serum enzyme levels in patients exposed to high doses of various drugs that affect the central nervous system. Increases in CK, AST, and alanine transanimase (ALT) activities were noted when the patients were comatose; however, their cardiac status was not reported in detail. Since isoenzyme evaluation was not included in this study, the ultimate source of the increased enzyme activity was not clear. Barbiturate and glutethimide poisoning have also been associated with increases in CK activity,[115–117] presumably due to hypothermia experienced during the coma. Increased CK activity, suggestive of myocardial damage, was detected in a child who accidentally ingested a large dose of imipramine.[118]

Ethanol

It is well known that dilated cardiomyopathy develops in a significant number of patients who consume large amounts of ethanol.[119] However, many details of this association remain to be clarified, particularly with respect to the pathogenesis of the cardiac damage that develops in these patients. A number of reports have described increased serum levels of myocardium-derived enzymes after the excessive intake of ethanol. In such patients, clinical evidence of cardiomyopathy has been accompanied by significant increases in serum CK,[120–123] CK-MB,[124] AST, and LDH.[121,122,125] In one study, the most profound increases in CK and CK-MB were found in patients who also had delirium tremens.[120,124] In a group of heavy drinkers with dilated cardiomyopathy, serum levels of CK, LDH, and AST were elevated in proportion to the amount of alcohol intake.[125] However, Osborn et al.[126] were not convinced of the diagnostic utility of CK, because they found that levels of this enzyme are commonly elevated in heavy users of alcohol who show no evidence of myocardial disease. These investigators concluded that CK-MB is a better indicator of alcohol-induced myocardial damage. Denison et al.[127] studied ECG changes and alterations in CK-MB, cTnT, and urinary catecholamines in alcohol-dependent men during withdrawal of alcoholic intake. Increases in CK-MB on the day of admission correlated with increases in urinary catecholamines. The study concluded that sympathoadrenergic activation could cause myocyte alterations and increases in serum levels of myocardial enzymes in susceptible individuals during withdrawal. It is clear, however, that ethanol-induced skeletal muscle damage must be considered as an alternative source of the elevation of these enzyme levels. This was found to be the case in a patient reported by Pottgen and Davis.[128] These observations emphasize the practical difficulties encountered in the evaluation of alcohol-induced cardiac damage by means of nonspecific biomarkers, since this agent usually affects multiple organ systems.

Toxins

The use of cardiac biomarkers has been applied to the assessment of myocardial injury after exposure to certain animal toxins. Scorpion envenomization in children can cause profound cardiovascular alterations (bradycardia, tachycardia, and hypertension), which, in some cases, can compromise myocardial function.[129] Abnormal ECG patterns and elevated ALT and CK levels were detected in a small number of hospitalized children who developed heart failure and pulmonary edema after scorpion bites.[129] Sofer et al.[130] attempted to assess the frequency of myocardial injuries in children after scorpion envenomization by measuring the plasma levels of the CK-MB isoenzyme. They identified 27 out of 32 poisoned children with signs of systemic intoxication (central nervous, cardiovascular, and gastrointestinal systems). Thirteen of these 27 children had elevated CK and CK-MB concentrations. They concluded that CK-MB determinations appeared to be more useful than other methods in detecting myocardial damage since only 5 of the 13 symptomatic children had ECG patterns that were consistent with myocardial injury. However, they found no relationship between levels of serum enzymes and the severity of intoxication.

Lalloo et al.[131] monitored the ECGs and cTnT levels in a series of 76 patients bitten by three types of New Guinea snakes (Taipans [69 patients], death adder [6 patients], and eastern brown [1 patient]). ECG changes were detected in 39 patients. Increases in cTnT (1 to 2.7 μg/L, compared with the normal reference range of <0.1 μg/L cTnT) were seen in only 3 of 39 patients tested both at the time of admission and approximately 24 hours after the bite (2 of 24 with ECG changes and 1 of the 15 with normal ECG patterns). These investigators concluded that myocardial damage, as indicated by increases in cTnT, is uncommon after envenomization by these three types of elapid snakes.

Okano et al.[132] evaluated a patient who had developed weakness and myalgia within 5 hours after eating the raw meat and liver from a blue humphead fish suspected of containing palytoxin. Detection of significant increases in serum concentrations of CK and myoglobin and in the urinary concentration of myoglobin were ascribed to rhabdomyolysis (a complication of palytoxin poisoning). In addition, the detection of increased serum levels of myosin light chain together with changes in the ECG were thought to indicate that myocardial injury might also have occurred in this patient. The authors note that this is the first instance in which myocardial injury associated with palytoxin poisoning was detected by assaying serum levels of myosin light chain. Thus, the use of diverse cardiac biomarkers should be extremely helpful in the systematic evaluation of the myocardial lesions produced by animal venoms and toxins.

Conclusions

Serum levels of specific proteins or enzymes that are released from damaged or necrotic cardiac myocytes can be used as biomarkers to detect toxic myocardial injury induced by a variety of substances in humans and experimental animals. The most extensively used cardiac biomarkers of toxic injury have been AST, LDH, LDH isoenzymes (LDH1), CK, and CK-MB. Recently, a great deal of interest has developed in the use of cTnT and cTnI, two components of the contractile system of the myocyte, for detecting toxic myocardial injury. The troponins are highly sensitive and specific markers of toxic injury because, in contrast to other proteins, they are immunohistochemically distinguishable from those derived from noncardiac tissue (skeletal muscle) and have a high degree of homology across a wide range of species. The assays for the troponins have been improved and standardized to a much greater degree for cTnT than for cTnI. Initial studies, using the troponins as biomarkers, have demonstrated the applicability of these methods to the study of the cardiotoxicity induced by DXR, an antineoplastic agent that causes a chronic dilated cardiomyopathy. The use of plasma ANP and BNP levels, which reflect cardiac dysfunction rather than direct myocyte injury, are also being evaluated as markers for DXR cardiotoxicity.

References

1. Billingham ME. Role of endomyocardial biopsy in diagnosis and treatment of heart disease. In Silver MD (ed): *Cardiovascular Pathology.* 2nd ed. New York: Churchill Livingstone; 1991:1465–1486.
2. Karmen A, Wroblewski F, La due JS. Transaminase activity in human blood. *Clin Invest* 1954;34:126–133.
3. La due JS, Wroblewski F, Karmen A. Serum glutamic oxaloacetic transaminase activity in human acute myocardial infarction science. *Science* 1954; 120:497–499.
4. Vessell ES, Beara AG. Localization of lactic acid dehydrogenase activity in blood. *Proc Soc Exp Biol Med* 1955;90:210–213.
5. Vasudevan G, Mercer DW, Varat MA. Lactic dehydrogenase isoenzyme determination in the diagnosis of acute myocardial infarction. *Am J Cardiol* 1978;57:1055–1057.
6. Dreyfus JC, Schapira G, Resnais J, et al. La creatine kinase serique dans la diagnostic de I=infarctus myocardique. *Rev Fr Etud Clin Biol* 1960;5: 386–387.
7. vander Veen KJ, Willebrand AF. Isoenzymes of creatine phosphokinase in tissue extracts and in normal and pathological serum. *Clin Chim Acta* 1966;13:312–316.
8. Roberts R, Gowda KS, Ludbrook PA, et al. Specificity of elevated serum MB creatine phosphokinase activity in the diagnosis of acute myocardial infarction. *Am J Cardiol* 1975;36:433–437.
9. Apple FS, Presse LM. Creatine kinase-MB: Detection of myocardial infarction and monitoring reperfusion. *J Clin Immunoassay* 1994;17:24–29.
10. Tsung JS, Tsung SS. Creatine kinase isoenzymes in extracts of various human skeletal muscles. *Clin Chem* 1986;32:1568–1570.
11. Chesebro MJ. Using serum markers in the early diagnosis of myocardial infarction. *Am Fam Physician* 1997;55:2667–2674.
12. Cummins B, Auckland ML, Cummins P. Cardiac-specific troponin-I radioimmunoassay in the diagnosis of acute myocardial infarction. *Am Heart J* 1987;113:1333–1344.

13. Katus HA, Remppis A, Neumann FJ, et al. Diagnostic efficiency of troponin T measurements in acute myocardial infarction. *Circulation* 1991;83: 902–912.
14. Perry SV. The regulation of contractile activity in muscle. *Biochem Soc Trans* 1979;7:593–617.
15. Wilkinson JM, Grand RJ. Comparison of amino acid sequence of troponin I from different striated muscles. *Nature* 1978;271:31–35.
16. Pearlstone JR, Carpenter MR, Smillie LB. Amino acid sequence of rabbit cardiac troponin T. *J Biol Chem* 1986;261:16795–16810.
17. Sasse S, Brand NJ, Kyprianou P, et al. Troponin I gene expression during human cardiac development and in end-stage heart failure. *Circ Res* 1993; 72:932–938.
18. Anderson PA, Malouf NN, Oakeley AE, et al. Troponin T isoform expression in humans. A comparison among normal and failing adult heart, fetal heart, and adult and fetal skeletal muscle. Circ Res 1991;69:1226–1233.
19. Muller-Bardorff M, Hallermayer K, Schroder A, et al. Improved troponin T ELISA specific for cardiac troponin T isoform: Assay development and analytical and clinical validation. *Clin Chem* 1997;43:458–466.
20. Saggin L, Gorza L, Ausoni S, et al. Cardiac troponin T in developing, regenerating and denervated rat skeletal muscle. *Development* 1990;110: 547–554.
21. Apple FS. Tissue specificity of cardiac troponin I, cardiac troponin T and creatine kinase-MB. *Clin Chim Acta* 1999;284:151–159.
22. Ricchiuti V, Voss EM, Ney A. Cardiac troponin T isoforms expressed in renal diseased skeletal muscle will not cause false-positive results by the second generation cardiac troponin T assay by Boehringer Mannheim. *Clin Chem* 1998;44:1919–1924.
23. Kampmann M, Raucher T, Mueller-Bardorff M, et al. Clinical evaluation of the cardiac markers troponin T and CK-MB on the Elecsys 2010 system. *Clin Chem* 1997;43:S159.
24. Katus HA, Remppis A, Scheffold T, et al. Intracellular compartmentation of cardiac troponin T and its release kinetics in patients with reperfused and nonreperfused myocardial infarction. *Am J Cardiol* 1991;67:1360–1367.
25. Mair J. Progress in myocardial damage detection: New biochemical markers for clinicians. *Crit Rev Clin Lab Sci* 1997;34:1–66.
26. Holt DW. Pre-clinical application of markers of myocardial damage. In Kaski JC, Holt DW (eds): *Myocardial Damage: Early Detection by Novel Biochemical Markers.* Boston: Kluwer Academic Publishers; 1998:201–211.
27. Malouf NN, McMahon D, Oakeley AE, Anderson PA. A cardiac troponin T epitope conserved across phyla. *J Biol Chem* 1992;267:9269–9274.
28. O'Brien PJ, Dameron GW, Beck ML, Brandt M. Differential reactivity of cardiac and skeletal muscle from various species in two generations of cardiac troponin-T immunoassays. *Res Vet Sci* 1998;65:135–137.
29. Remppis A, Schefford T, Greten J, et al. Intracellular compartmentation of troponin T: Release kinetics after global ischemia and calcium paradox in the isolated perfused rat heart. *J Mol Cell Cardiol* 1995;27:793–803.
30. Voss EM, Sharkey SW, Gernert AE, et al. Human and canine cardiac troponin T and creatine kinase-MB distribution in normal and diseased myocardium. Infarct sizing using serum profiles. *Arch Pathol Lab Med* 1995;119: 799–806.
31. O'Brien PJ, Dameron GW, Beck ML, et al. Cardiac troponin T is a sensitive, specific biomarker of cardiac injury in laboratory animals. *Lab Anim Sci* 1997;47:486–495.

32. Walpoth BH, Tschopp A, Peheim E, et al. Assessment of troponin-T for detection of cardiac rejection in a rat model. *Transplant Proc* 1995;27: 2084–2087.
33. Bachmaier K, Mair J, Offner F, et al. Serum cardiac troponin T and creatine kinase-MB elevations in murine autoimmune myocarditis. *Circulation* 1995;92:1927–1932.
34. Smith SC, Ladenson JH, Mason JW, Jaffe AS. Elevations of cardiac troponin I associated with myocarditis. Experimental and clinical correlates. *Circulation* 1997;95:163–168.
35. Bleuel H, Deschl U, Bertsch T, et al. Diagnostic efficiency of troponin T measurements in rats with experimental myocardial cell damage. *Exp Toxicol Pathol* 1995;47:121–127.
36. Yandle T, Crozier I, Nicholls G, et al. Amino acid sequence of atrial natriuretic peptides in human coronary sinus plasma. *Biochem Biophys Res Commun* 1987;146:832–839.
37. Holmes SJ, Espiner BA, Richards AM, et al. Renal, endocrine, and hemodynamic effects of human brain natriuretic peptide in normal man. *J Clin Endocrinol Metab* 1993;76:91–96.
38. Sagnella GA. Measurement and significance of circulating natriuretic peptides in cardiovascular disease. *Clin Sci (Colch)* 1998;95:519–529.
39. Mukoyama M, Nakao K, Hosoda K, et al. Brain natriuretic peptide as a novel cardiac hormone in humans. Evidence for an exquisite dual natriuretic peptide system, atrial natriuretic peptide and brain natriuretic peptide. *J Clin Invest* 1991;87:1402–1412.
40. Levin ER, Gardner DG, Samson WK. Natriuretic peptides. *N Engl J Med* 1998;339:321–328.
41. Sundsfjord JA, Thibault G, Larochelle P, Cantin M. Identification and plasma concentrations of the N-terminal fragment of proatrial natriuretic factor in man. *J Clin Endocrinol Metab* 1988;66:605–610.
42. Takeda T, Kohno M. Brain natriuretic peptide in hypertension. *Hypertens Res* 1995;18:259–266.
43. Bevilacqua M, Vago T, Baldi G, et al. Analytical agreement and clinical correlates of plasma brain natriuretic peptide measured by three immunoassays in patients with heart failure. *Clin Chem* 1997;43:2439–2440.
44. Young RC, Ozols RF, Myers CE. The anthracycline antineoplastic drugs. *N Engl J Med* 1981;305:139–153.
45. Ganz WI, Sridhar KS, Ganz SS, et al. Review of tests for monitoring doxorubicin-induced cardiomyopathy. *Oncology* 1996;53:461–470.
46. Olson HM, Shannon CF. Alterations of lactate dehydrogenase isoenzymes with subacute adriamycin toxicity. *Cancer Treat Rep* 1979;63:2057–2059.
47. Bhanumathi P, Saleesh EB, Vasudevan DM. Creatine phosphokinase and cardiotoxicity in adriamycin chemotherapy and its modification by WR-1065. *Biochem Arch* 1992;8:335–338.
48. Dobric S, Dragojevic-Simic V, Bokonjic D, et al. The efficacy of selenium, WR-2721, and their combination in the prevention of adriamycin-induced cardiotoxicity in rats. *J Environ Pathol Toxicol Oncol* 1998;17:291–299.
49. De Leonardis V, Neri B, Bacalli S, Cinelli P. Reduction of cardiac toxicity of anthracyclines by L-carnitine: Preliminary overview of clinical data. *Int J Clin Pharmacol Res* 1985;5:137–142.
50. Fink FM, Genser N, Fink C, et al. Cardiac troponin T and creatine kinase MB mass concentrations in children receiving anthracycline chemotherapy. *Med Pediatr Oncol* 1995;25:185–189.

51. Missov E, Calzolari C, Davy JM, et al. Cardiac troponin I in patients with hematologic malignancies. *Coron Artery Dis* 1997;8:537–541.
52. Seino Y, Tomita Y, Nagai Y, et al. Cardioprotective effects of ace-inhibitor (Cilazapril) on adriamycin cardiotoxicity in spontaneously hypertensive rats. *Circulation* 1993;88:I633. Abstract.
53. Raderer M, Kornek G, Weinlander G, Kastner J. Serum troponin T levels in adults undergoing anthracycline therapy. *J Natl Cancer Inst* 1997;89:171.
54. Ottinger ME, Sallan S, Rifai N, et al. Myocardial damage in doxorubicin-treated children: A study of serum cardiac troponin T. *Proc Am Soc Clin Oncol* 1995;14:345. Abstract.
55. Lipshultz SE, Rifai N, Sallan SE, et al. Predictive value of cardiac troponin T in pediatric patients at risk for myocardial injury. *Circulation* 1997;96: 2641–2648.
56. Herman EH, Lipshultz SE, Rifai N, et al. Use of cardiac troponin T levels as an indicator of doxorubicin-induced cardiotoxicity. *Cancer Res* 1998;58: 195–197.
57. Herman EH, Zhang J, Lipshultz SE, et al. Correlation between serum levels of cardiac troponin-T and the severity of the chronic cardiomyopathy induced by doxorubicin. *J Clin Oncol* 1999;17:2237–2243.
58. Adamcova M, Gersl V, Hrdina R, et al. Cardiac troponin T as a marker of myocardial damage caused by antineoplastic drugs in rabbits. *J Cancer Res Clin Oncol* 1999;125:268–274.
59. Bauch M, Ester A, Kimura B, et al. Atrial natriuretic peptide as a marker for doxorubicin-induced cardiotoxic effects. *Cancer* 1992;69:1492–1497.
60. Tikanoja T, Riikonen P, Perkkio M, Helenius T. Serum N-terminal atrial natriuretic peptide (NT-ANP) in the cardiac follow-up in children with cancer. *Med Pediatr Oncol* 1998;31:73–78.
61. Sothern RB, Vesely DL, Kanabrocki EL, et al. Temporal (circadian) and functional relationship between atrial natriuretic peptides and blood pressure. *Chronobiol Int* 1995;12:106–120.
62. Toyoda Y, Okada M, Kashem MA. A canine model of dilated cardiomyopathy induced by repetitive intracoronary doxorubicin administration. *J Thorac Cardiovasc Surg* 1998;115:1367–1373.
63. Yokota N, Yamamoto Y, Iemura F, et al. Increased plasma levels and effects of brain natriuretic peptide in experimental nephrosis. *Nephron* 1993;65:454–459.
64. Suzuki T, Hayashi D, Yamazaki T, et al. Elevated B-type natriuretic peptide levels after anthracycline administration. *Am Heart J* 1998;136: 362–363.
65. Nousiainen T, Jantunen E, Vanninen E, et al. Natriuretic peptides as markers of cardiotoxicity during doxorubicin treatment for non-Hodgkin's lymphoma. *Eur J Haematol* 1999;62:135–141.
66. Yamashita J, Ogawa M, Shirakusa T. Plasma endothelin-1 as a marker for doxorubicin cardiotoxicity. *Int J Cancer* 1995;62:542–547.
67. Wei CM, Lerman A, Rodeheffer RJ, et al. Endothelin in human congestive heart failure. *Circulation* 1994;89:1580–1586.
68. Labianca R, Beretta G, Clerici M, et al. Cardiac toxicity of 5–fluorouracil: A study of 1083 patients. *Tumori* 1982;68:505–510.
69. Braumann D, Mainzer K, Gunzl C, Lewerenz B. Myokardinfarkte im Rahmen einer 5-Fluorouracil-Therapie. *Onkologie* 1990;13:465–467.
70. Cersosimo RS, Lee JM. Creatine kinase elevation associated with 5-fluorouracil and levamisole therapy for carcinoma of the colon. A case report. *Cancer* 1996;77:1250–1253.

71. Pan L, Yang X, Song H. [Cardiotoxicity of 5-fluorouracil]. *Chung Hua Fu Chan Ko Tsa Chih* 1996;31:86–89.
72. Schmidt JL, Fleming WW. The structure of sympathomimetics as related to reserpine induced sensitivity changes in the rabbit ileum. *J Pharmacol Exp Ther* 1962;139:230–237.
73. Dilsaver SC, Votolato NA, Alessi NE. Complications of phenylpropanolamine. *Am Fam Physician* 1989;39:201–206.
74. Pentel PR, Mikell FL, Zavoral JH. Myocardial injury after phenylpropanolamine ingestion. *Br Heart J* 1982;47:51–54.
75. Leo PJ, Hollander JE, Shih RD, Marcus SM. Phenylpropanolamine and associated myocardial injury. *Ann Emerg Med* 1996;28:359–362.
76. Wiener I, Tilkian AG, Palazzolo M. Coronary artery spasm and myocardial infarction in a patient with normal coronary arteries: Temporal relationship to pseudoephedrine ingestion. *Cathet Cardiovasc Diagn* 1990;20:51–53.
77. Meltzer Y, Guschwan A. Effect of isoproterenol on rat plasma creatine phosphokinase activity. *Arch Int Pharmacodyn Ther* 1971;194:141–146.
78. Bernauer W. Effect of isoproterenol on myocardial creatine kinase activity in rats. *Arch Int Pharmacodyn Ther* 1978;231:90–97.
79. Tsuboi T, Ishikawa K, Ohsawa Y, et al. Biochemical changes of myocardial necrosis induced by isoproterenol and protective effects of beta-blocker and anti-inflammatory drugs. *Chem Pharm Bull (Tokyo)* 1974;22:669–675.
80. Preus M, Bhargava AS, Khater AE, Gunzel P. Diagnostic value of serum creatine kinase and lactate dehydrogenase isoenzyme determinations for monitoring early cardiac damage in rats. *Toxicol Lett* 1988;42:225–233.
81. Bertsch T, Bleuel H, Aufenanger J, Rebel W. Comparison of cardiac troponin T and cardiac troponin I concentrations in peripheral blood during orciprenaline induced tachycardia in rats. *Exp Toxicol Pathol* 1997;49: 467–468.
82. Bertsch T, Bleuel H, Deschl U, Rebel W. A new sensitive cardiac troponin T rapid test (TROPT) for the detection of experimental acute myocardial damage in rats. *Exp Toxicol Pathol* 1999;51:565–569.
83. Benedetti TJ. Maternal complications of parenteral beta-sympathomimetic therapy for premature labor. *Am J Obstet Gynecol* 1983;145:1–6.
84. Meinen K, Breinl H, Schmidt EW, Wellstein A. Necroses du myocarde par tocolyse avec le fenoterol? *J Gynecol Obstet Biol Reprod (Paris)* 1979;8: 23–25.
85. Steyer M, Rink K, Schlesing H, et al. Serum kreatinkinase BM wahrend Fenoterol-Tokolyse. *Z Geburtshilfe Perinatol* 1979;183:339–342.
86. Meinen K, Breinl H. Creatinkinase-B-Radioimmunoassay—Seine Bedeutung bei der Fahndung nach Myokardnekrosen unter Langzeittokolyse mit Fenoterol. *Z Geburtshilfe Perinatol* 1980;184:339–342.
87. Meinen K, Breinl H, Leuterer W. Etat cardiaque des nouveau-nes apres une tocolyse prolongee avec le fenoterol. *J Gynecol Obstet Biol Reprod (Paris)* 1983;12:315–320.
88. Hadi HA, Albazzaz SJ. Cardiac isoenzymes and electrocardiographic changes during ritodrine tocolysis. *Am J Obstet Gynecol* 1989;161:318–321.
89. Gerris J, Bracke M, Thiery M, Van Maele G. Cardiotoxicity of ritodrine: Assessment based on serum creatine kinase activity. *Z Geburtshilfe Perinatol* 1980;184:25–30.
90. Leiserowitz GS, Evans AT, Samuels SJ, et al. Creatine kinase and its MB isoenzyme in the third trimester and the peripartum period. *J Reprod Med* 1992;37:910–916.

91. Adamcova M, Kokstein Z, Palicka V, et al. Hladiny troponinu T v pupecnikove krvi po tokolyze Bricanylem. *Ceska Gynekol* 1996;61:210–214.
92. Adamcova M, Kokstein Z, Palicka V, et al. Cardiac troponin T in pregnant women having intravenous tocolytic therapy. *Arch Gynecol Obstet* 1999; 262:121–126.
93. Sykes AP, Lawson N, Finnegan JA, Ayres JG. Creatine kinase activity in patients with brittle asthma treated with long term subcutaneous terbutaline. *Thorax* 1991;46:580–583.
94. Stephanopoulos DE, Monge R, Schell KH, et al. Continuous intravenous terbutaline for pediatric status asthmaticus. *Crit Care Med* 1998;26: 1744–1748.
95. Strubelt O, Hoffmann A, Siegers CP, Sierra-Callejas JL. On the pathogenesis of cardiac necroses induced by theophylline and caffeine. *Acta Pharmacol Toxicol (Copenh)* 1976;39:383–392.
96. Lindamood C III, Lamb JC IV, Bristol DW, et al. Studies on the short-term toxicity of theophylline in rats and mice. *Fundam Appl Toxicol* 1988;10: 477–489.
97. Watson ES, Griswold DE, Schwartz LW, et al. Theophylline induces splanchnic arterial damage in rats. *Toxicol Sci* 1999;48:299.
98. Delahunty TJ. Cathodally migrating creatine kinase in sera of patients treated with theophylline, and its diagnostic implications. *Clin Chem* 1983; 29:1484–1487.
99. Ng RH, Roe C, Funt D, Statland BE. Increased activity of creatine kinase isoenzyme MB in a theophylline-intoxicated patient. *Clin Chem* 1985;31: 1741–1742.
100. Chazan R, Karwat K, Tyminska K, et al. Wplyw dozylnych wlewow teofiliny na wystepowanie zaburzen rytmu serca i aktywnosc frakcji sercowej kinazy kreatynowej CK-MB u chorych z obturacja drog oddechowych. *Pol Arch Med Wewn* 1993;90:399–408.
101. Chazan R, Karwat K, Tyminska K, et al. Cardiac arrhythmias as a result of intravenous infusions of theophylline in patients with airway obstruction. *Int J Clin Pharmacol Ther* 1995;33:170–175.
102. Coleman DL, Ross TF, Naughton JL. Myocardial ischemia and infarction related to recreational cocaine use. *West J Med* 1982;136:444–446.
103. Goldfrank LR, Hoffman RS. The cardiovascular effects of cocaine. *Ann Emerg Med* 1991;20:165–175.
104. Brody SL, Slovis CM, Wrenn KD. Cocaine-related medical problems: Consecutive series of 233 patients. *Am J Med* 1990;88:325–331.
105. Hollander JE, Hoffman RS, Gennis P, et al. Prospective multicenter evaluation of cocaine-associated chest pain. Cocaine Associated Chest Pain (COCHPA) Study Group. *Acad Emerg Med* 1994;1:330–339.
106. Gitter MJ, Goldsmith SR, Dunbar DN, Sharkey SW. Cocaine and chest pain: Clinical features and outcome of patients hospitalized to rule out myocardial infarction. *Ann Intern Med* 1991;115:277–282.
107. Tokarski GF, Paganussi P, Urbanski R, et al. An evaluation of cocaine-induced chest pain. *Ann Emerg Med* 1990;19:1088–1092.
108. Amin M, Gabelman G, Karpel J, Buttrick P. Acute myocardial infarction and chest pain syndromes after cocaine use. *Am J Cardiol* 1990;66: 1434–1437.
109. Rubin RB, Neugarten J. Cocaine-induced rhabdomyolysis masquerading as myocardial ischemia. *Am J Med* 1989;86:551–553.
110. Roth D, Alarcon FJ, Fernandez JA, et al. Acute rhabdomyolysis associated with cocaine intoxication. *N Engl J Med* 1988;319:673–677.

111. McLaurin M, Henry TD, Apple FS, Sharkey SW. Cardiac troponin I, T and CK-MB in patients with cocaine-related chest pain. *Circulation* 1994; 90:I278
112. McLaurin M, Apple FS, Henry TD. Cardiac troponin I and T concentrations in patients with cocaine associated chest pain. *Ann Clin Biochem* 1996; 33:183–186.
113. Hollander JE, Levitt MA, Young GP, et al. Effect of recent cocaine use on the specificity of cardiac markers for diagnosis of acute myocardial infarction. *Am Heart J* 1998;135:245–252.
114. Wright N, Clarkson AR, Brown SS, Fuster V. Effects of poisoning on serum enzyme activities, coagulation, and fibrinolysis. *Br Med J* 1971;3:347–350.
115. Henderson LW, Metz M, Wilkinson JH. Serum enzyme elevation in glutethimide intoxication. *Br Med J* 1970;3:751.
116. Smith AF, Radford D, Wong CP, Oliver MF. Creatine kinase MB isoenzyme studies in diagnosis of myocardial infarction. *Br Heart J* 1976;38: 225–232.
117. Sabiniewicz M, Szajawski JM. Creatine phosphokinase and barbiturate overdosage. *Lancet* 1979;1:152.
118. Sueblinvong V, Wilson JF. Myocardial damage due to imipramine intoxication. *J Pediatr* 1969;74:475–478.
119. Regan TJ, Ettinger PO, Haider B, et al. The role of ethanol in cardiac disease. *Annu Rev Med* 1977;28:393–409.
120. Lakair JS, Myerson RM. Alcoholic myopathy. *Arch Intern Med* 1968;122: 417–422.
121. Curran JR, Wetmore SJ. Alcoholic myopathy. *Dis Nerv Syst* 1972;33:19–22.
122. Spector R, Choudhury A, Cancilla P, Lakin R. Alcohol myopathy. Diagnosis by alcohol challenge. *JAMA* 1979;242:1648–1649.
123. Nygren A. Serum creatine phosphokinase activity in chronic alcoholism, in connection with acute alcohol intoxication. *Acta Med Scand* 1966;179: 623–630.
124. Chemnitz G, Schmidt E, Schmidt FW, Lobers J. Das Creatinkinase Isoenzym MB im Serum bei extrakardialen Erkrankungen. *Verh Dtsch Ges Inn Med* 1978;1587–1590.
125. Richardson PJ, Wodak AD, Atkinson L. Relation between alcohol intake, myocardial enzyme activity, and myocardial function in dilated cardiomyopathy. Evidence for the concept of alcohol induced heart muscle disease. *Br Heart J* 1986;56:165–170.
126. Osborn LA, Rossum A, Standefer J, et al. Evaluation of CK and CK-MB in alcohol abuse subjects with recent heavy consumption. *Cardiology* 1995; 86:130–134.
127. Denison A, Jern S, Jagenburg B. ST-segment changes and catecholamine related myocardial enzyme release during alcohol withdraw. *Alcohol* 1997; 32:185–194.
128. Pottgen P, Davis ER. Increased creatine kinase MB values in acute alcoholic intoxication. *Clin Chem* 1980;26:792.
129. Gueron M, Yaron R. Cardiovascular manifestations of severe scorpion sting. Clinicopathologic correlations. *Chest* 1970;57:156–162.
130. Sofer S, Shahak E, Slonim A, Gueron M. Myocardial injury without heart failure following envenomation by the scorpion Leiurus quinquestriatus in children. *Toxicon* 1991;29:382–385.

131. Lalloo DG, Trevett AJ, Nwokolo N, et al. Electrocardiographic abnormalities in patients bitten by taipans (Oxyuranus scutellatus canni) and other elapid snakes in Papua New Guinea. *Trans R Soc Trop Med Hyg* 1997;91: 53–56.
132. Okano H, Masuoka H, Kamei S, et al. Rhabdomyolysis and myocardial damage induced by palytoxin, a toxin of blue humphead parrotfish. *Intern Med* 1998;37:330–333.

Chapter 20

The Utility of Brain Natriuretic Peptides in Patients with Heart Failure and Coronary Artery Disease

Johannes Mair, MD

Introduction: Prevalence, Incidence, and Significance of Heart Failure and Left Ventricular Dysfunction

Heart failure is a significant clinical problem, particularly in the elderly, and is one of the main causes of hospitalization in industrialized countries. The most common causes of heart failure in industrialized countries are coronary artery disease (CAD) and hypertension.[1,2] Even though people on the average are getting older, major improvements in the treatment of acute myocardial infarction (AMI) could markedly reduce the acute mortality of heart attacks although the survivors would sustain the long-term consequences of heart failure. The prevalence of heart failure is higher in men than in women. The prevalence and incidence of the disease almost doubled during the last 10 years, and both increase with age. The prevalence of heart failure in people over 70 years of age is approximately 10% to 15%, and in these people up to 70% of hospital admissions are directly or indirectly related to heart failure. The incidence of heart failure is age-dependent; in people aged 45 to 74 years the average incidence is estimated to be 4% per year.[3] The overall prevalences of echocardiographically defined left ventricular dysfunction (LVD) in two population-based studies were 3.7% (Rotterdam Study) and 2.9% (Glasgow MONICA risk factor survey), with a marked increase with age observed in both studies (Table 1).[4,5] LVD was asymptomatic in approximately half of the cases.

From: Adams JE III, Apple FS, Jaffe AS, Wu AHB (eds). *Markers in Cardiology: Current and Future Clinical Applications.* Armonk, NY: Futura Publishing Company, Inc.; © 2001.

Table 1

Prevalence of Heart Failure in Two Population-Based Studies

Age (Years)	Rotterdam Study			General Practice Study, UK*
	Men	Women	All	All
55–64	0.7%	0.6%	0.7%	
65–74	3.7%	1.6%	2.7%	2.7%
75–84	14.4%	12.1%	13.0%	7.4%

* Morbidity statistics from general practice. Fourth National Survey, 1991–92. Royal College of General Practitioners, Office of Population Census and Survey, and Department of Health and Social Security, London: HMSO, 1995.

This is the reason for the so-called iceberg phenomenon of heart failure, which means that the majority of patients with heart failure and LVD go undiagnosed and thus without adequate treatment.

Most patients with heart failure are managed by general practitioners, and only a minority are seen by cardiologists. However, timely recognition and therapeutic intervention are essential to improve quality of life and prognosis, because it has been demonstrated that by treatment with angiotensin-converting enzyme (ACE) inhibitors, the progression to clinically manifest heart failure can be slowed.[6] The diagnosis of heart failure, however, is fraught with difficulties because clinical signs and symptoms have limited sensitivities and specificities and only allow for a rather imprecise diagnosis in the early stages of heart failure.[7] The symptoms are not specific and the clinical signs, although reasonably specific, are not at all sensitive. Consequently, even experienced physicians disagree on the diagnosis, especially when heart failure is mild. Only about half of the patients with a primary care diagnosis of heart failure have evidence of this disorder on further cardiological assessment. Heart failure is a diagnostic challenge particularly in the elderly, in whom clinical signs and symptoms (eg, exertional dyspnea, fatigue, peripheral edema) have reduced specificity, and comorbidity is common (eg, respiratory diseases and physical deconditioning).

Heart failure is associated with significant morbidity and mortality.[8] Heart failure markedly impairs the quality of life, particularly in patients with moderate and severe heart failure. The 5-year mortality in heart failure patients is approximately 75% in men and 60% in women; the 1-year mortality in end-stage disease (New York Heart Association [NYHA] Class IV) is up to 50% to 60%. Obviously, heart failure is a significant burden for the individual patient and a major socioeconomic burden for the community. Consequently, the early identification of LVD can have a major impact on patients and on society. The current, common consensus is that therapy should be started in all symptomatic patients or asymptomatic patients with a left ventricular ejection fraction (LVEF) below 40%.[6] The goals are to improve symptoms, to delay disease progression, and to reduce morbidity and mortality. The timely initiation of ACE inhibitor treatment is a cornerstone to achieving these goals, which requires reliable early diagnosis of LVD.[6] Consequently, a sensitive and specific laboratory marker of early heart failure would be of obvious significant clinical benefit. The test should reliably allow for ruling out heart failure and should have an adequate positive predictive value.

Pathophysiology of Cardiac Natriuretic Peptides

Approximately 20 years ago, de Bold and coworkers[9] demonstrated the existence of the long-predicted natriuretic factor and the endocrine function of the heart. They reported on a hormonal link between the heart and the kidneys based on the intravenous administration of atrial extracts into rats that resulted in a marked increase in sodium excretion. In the meantime, the endocrine function of the heart became well understood, and this discovery finally led to the characterization of a whole family of structurally similar but genetically distinct peptides, the natriuretic peptides (NPs).[10–12] The precursor prohormones for each peptide are encoded by separate genes, and the tissue distribution and regulation of each peptide are unique. NPs are the main hormones in the body's defense against volume overload and hypertension. Atrial NP (ANP) and brain NP (BNP) are released in response to volume expansion and pressure overload of the heart, and thus they are broadly involved in the regulation of blood pressure, blood volume, and sodium balance (Table 2). NPs are the naturally

Table 2

Cardiovascular Effects of Natriuretic Peptides

1. Decrease of systemic vascular resistance (afterload) and systemic arterial pressure
2. Decrease of pulmonary vascular resistance and pulmonary arterial pressure
3. Venodilation and decrease of venous return with reduction of preload
4. Decrease of right atrial and left ventricular end-diastolic pressures
5. Coronary artery dilatation with decrease of coronary artery resistance and increase in coronary blood flow, but no direct positive inotropic or chronotropic effects
6. Increase in renal blood flow, glomerular filtration rate, and urine output

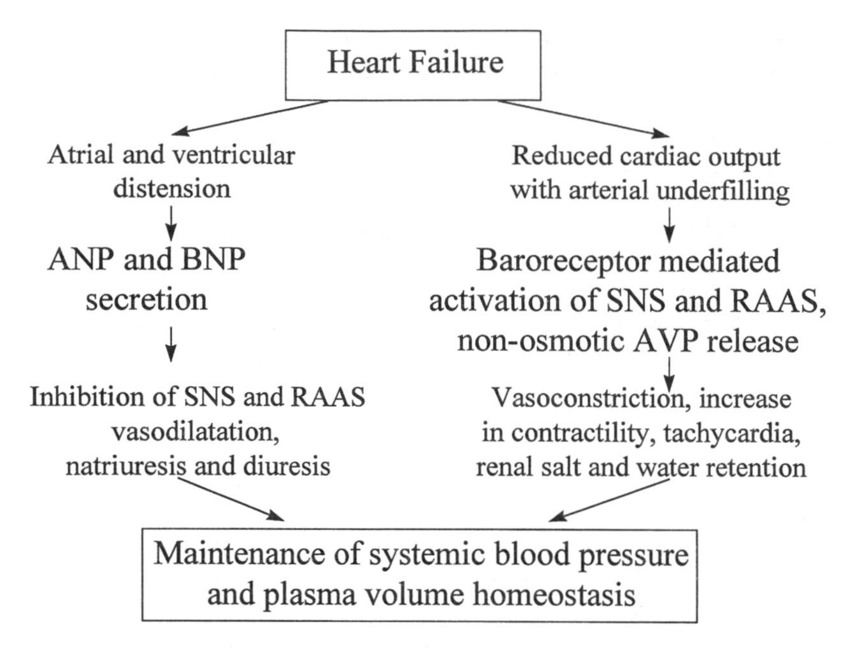

Figure 1. Schema of the complex neurohormonal interplay in heart failure. ANP = atrial natriuretic peptide; AVP = arginine vasopressin; BNP = brain natriuretic peptide; RAAS = renin-angiotensin-aldosterone system; SNS = sympathetic nervous system.

occurring antagonists of the hormones of the renin-angiotensin-aldosterone system (RAAS) and the sympathetic nervous system (SNS), peripherally as well as in the central nervous system (Fig. 1). NPs control fluid and electrolyte homeostasis via coordinated central and peripheral actions. All three NPs, particularly C-type NP (CNP), are produced in the brain, and the actions of NPs in the brain reinforce those in the periphery. The RAAS and SNS promote sodium and fluid balance and support blood pressure. NPs, as the physiological counterparts of the RAAS and SNS, oppose these systems in situations involving salt and fluid overload or elevated blood pressure. NPs inhibit the secretion of arginine vasopressin and corticotropin, and inhibit salt appetite and water drinking as well as sympathetic tone in the central nervous system.[13] Peripherally, NPs increase glomerular filtration rate and natriuresis to protect the heart from acute volume loads. The hemodynamic effects of NP consist of decreases in systemic and pulmonary vascular resistance, systemic and pulmonary arterial pressure, plasma volume, venous return pressure, right atrial pressure, pulmonary capillary wedge pressure, and left ventricular end-diastolic pressure.[14] As a result, the cardiac output increases and diastolic function improves. The rise of pulmonary capillary wedge pressure during exercise is attenuated. NPs are not endothelial-cell-dependent vasodilators. They dilate coronary arteries and increase coronary blood flow, but they do not have direct positive inotropic or chronotropic effects. However, secondary to the other effects of NPs, they improve myocardial performance and increase cardiac output in heart failure patients. In addition, NPs have antimitogenic activity in the cardiovascular system. The NPs potentially limit the myocardial proliferative or hypertrophic response to damage and the remodeling of vessels.

The NP family members relevant to cardiovascular diseases are ANP, which is also called A-type NP, BNP, also called B-type NP, and CNP, which, as previously mentioned, is also called C-type NP. The name "brain natriuretic peptide" is misleading because circulating BNP originates from the heart, and the highest concentrations of this peptide are found in myocardium. BNP, however, was first isolated from porcine brain tissue, which explains the name.[15] But both circulating ANP and BNP are of myocardial origin. CNP is of endothelial and renal epithelial origin. These three peptides are all characterized by a common 17 amino acid ring structure with a disulfide bond between two cysteine residues and high homology between different peptides (approximately 77% homology between human ANP and BNP [Fig. 2]). This ring structure is essential for receptor binding, and the disruption of this loop leads to loss of physiological activity. Whereas ANP

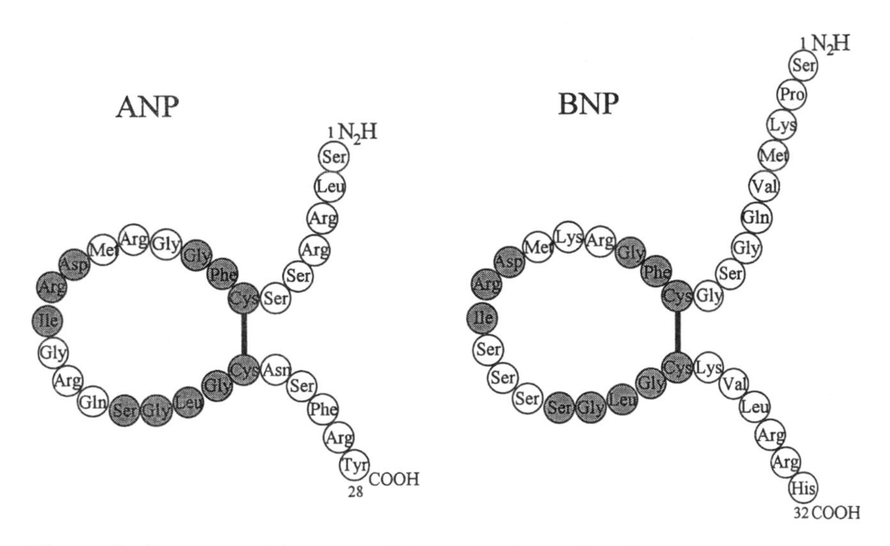

Figure 2. Structure of human atrial and brain natriuretic peptides (ANP and BNP, respectively). Homologous amino acids of both peptides in the ring structure, which is essential for receptor binding, are plotted in gray.

and BNP are mainly circulating hormones with natriuretic, vasodilatory, renin inhibitory, antimitotic, and lusitropic properties (Table 2), CNP acts as a paracrine factor with vasorelaxant and antiproliferative effects on vascular tissue but without natriuretic actions.[16] CNP molecules of 22 and 53 amino acids in length have been identified, but 22 amino acid CNP is the predominant CNP form. CNP is usually low or undetectable in plasma. In contrast to ANP and BNP, CNP is not increased in heart failure patients and does not increase in response to physical exercise.

Guanylyl cyclase-linked NP receptors (NPRs) mediate the known biological effects of NP and have cytoplasmic guanylyl cyclase domains that are stimulated when the receptors bind a ligand. The second messenger is cyclic guanosine monophosphate (cGMP). cGMP exerts its biological effects indirectly through cGMP-dependent protein kinase G or phosphodiesterases, or by direct actions on ion channels (for example, in the kidneys). NPR-A and NPR-B have guanylyl cyclase activity. NPR-A responds to ANP and, to a 10-fold lesser degree, to BNP. NPR-B responds primarily to CNP, but it also binds ANP and BNP, but binds BNP with higher affinity than that of ANP. No specific BNP receptor has been identified so far. NPR-A receptor is strongly expressed in the endothelium of vasculature (most abundant in large blood vessels), kidneys, and adrenal glands. NPR-A stimulation mediates vasorelaxant and natriuretic functions and decreases aldosterone synthesis. NPR-B receptor is strongly expressed in the brain, but it is also found in vascular smooth muscle. NPR-C is a third receptor type that has no guanylyl cyclase activity and does not mediate the known physiological effects of NP. NPR-C is thought to play an essential role in the removal of NP from the circulation and in the regulation of plasma NP concentrations, thus acting as a hormonal buffer system for impending, large, inappropriate fluctuations in plasma concentrations. The hormone receptor complex is internalized, the hormone subjected to lysosomal proteolysis, and the receptors subsequently recycled. NPR-C interacts with all three NPs in the order ANP > CNP > BNP. The clearance receptor binds ANP with a higher affinity than that of BNP, a factor that contributes to the longer biological half-life of BNP. NPR-C is the most widely and abundantly expressed NPR, with a distribution that includes most tissues that express NPR-A or NPR-B. More than 95% of all receptors in kidneys and vascular tissue are clearance receptors. NPR-C modulates the availability of NPs at their target organs, thereby tailoring the activity of the NP to specific local needs.[17] The elements of the system (ligand, active receptors, and clearance receptors) can be regulated independently.

Another mechanism that clears NP is enzymatic degradation by the ectoenzyme neutral endopeptidase 24.11 (NEP), which is widely expressed in the kidneys, lungs, and the vascular wall. Particularly high concentrations of this enzyme are found in the lung and kidneys. CNP has the highest affinity to this enzyme, followed by ANP and BNP. The affinity of ANP is much higher than that of BNP, which is another factor responsible for the longer half-life of BNP (ANP approximately 3 minutes versus BNP 18 minutes). Nonetheless, NEP-mediated degradation appears to be the predominant metabolic pathway of BNP as well.

Pro-ANP messenger ribonucleic acid is most abundant in the atria of the heart, where ANP is primarily produced. In LVD and ventricular hypertrophy, ANP is also derived from ventricular tissue.[18] Increased atrial wall tension, reflecting increased intravascular volume, is the dominant stimulus for ANP release. The primary gene product is a precursor called prepro-ANP, which consists of 151 amino acids in humans and has a 25 amino acid signal peptide sequence at its amino-terminal end. The prohormone pro-ANP (126 amino acids, also called γANP) is stored in granules within atrial cardiomyocytes after cleavage of the signal peptide. Regulation of ANP release occurs mainly at the level of hormone secretion. ANP is primarily a measure of extracellular volume status with very rapid response. Upon secretion, pro-ANP is split by a membrane bound protease (atriopeptidase) into a C-terminal and N-terminal portion in an equimolar basis. Both fragments enter the circulation. The C-terminal ANP (ANP or α-ANP) consists of 28 amino acids and is the physiologically active hormone. N-terminal pro-ANP (NT pro-ANP) occurs mainly in the high molecular mass form NT pro-ANP 1–98. NT pro-ANP may be cleaved into smaller fragments (pro-ANP 1–30, 31–67, and 79–98) with natriuretic, diuretic, or vasodilatory effects, although these fragments are not capable of binding to the specific NP receptors. The presence of smaller fragments of NT pro-ANP in human plasma has been reported by some investigators, but could not be confirmed by all investigators. In patients with heart failure, an antiparallel dimer of α-ANP, β-ANP, has been detected in myocardium and plasma.[19] β-ANP is a postprocessing product of mature α-ANP and has reduced physiological effects compared with α-ANP.

In contrast to pro-ANP, pro-BNP (108 amino acids) is not stored in granules. BNP is also secreted by the atria, but the greater proportion of circulating BNP comes from the ventricles. Acute regulation of BNP synthesis and excretion occurs at the level of gene expression. BNP is primarily a marker of systolic and diastolic cardiac dysfunction and cardiac hypertrophy; it is the slower reacting NP. Whether pro-BNP is split into BNP-32 and NT pro-BNP 1–76 upon secretion or pro-BNP is split later on in serum into the physiologically active hormone and the N-terminal portion is currently not definitively known. The presence of pro-BNP and BNP-32 in cardiomyocytes has been reported. In addition to BNP and NT pro-BNP, a high molecular mass peptide presumed to be the intact pro-BNP circulates in human plasma. The functional role of NT pro-BNP is also currently not exactly known. The major origin of BNP is ventricles. Left ventricular stretch and wall tension are the primary regulators of BNP release.

ANP and BNP form a dual, integrated NP system, and BNP may be a back-up hormone activated only after prolonged ventricular overload. Depending on the predominance of atrial or ventricular overload in heart failure patients, the secretion pattern of ANP and BNP varies because BNP predominantly reflects the degree of ventricular overload and ANP reflects the degree of atrial overload. In certain heart failure patients, BNP plasma concentrations may exceed ANP concentrations (for example, left ventricular, hypertrophic, obstructive cardiomyopathy).

Diseases Associated with Increased Plasma Concentrations of NPs

NPs are increased in various cardiac disorders such as left ventricular systolic or diastolic dysfunction, left ventricular hypertrophy, cor pulmonale, or myocarditis (Table 3). NPs are usually increased in the setting of systemic and pulmonary hypertension and are involved in preventing the development of systemic and pulmonary hypertension. NPs have a role in the defense against elevated blood pressure, particularly against mineralocorticoid-induced and salt-induced hypertensions. NPs lower blood pressure by stimulating natriuresis, by vasodilatation, by inhibiting sympathetic tone, and by the production and action of vasoconstrictor peptides.

Table 3

Diseases Associated with Increases in Natriuretic Peptides

1. Acute or chronic systolic or diastolic heart failure
2. Left ventricular hypertrophy
3. Inflammatory cardiac diseases (e.g., myocarditis, cardiac allograft rejection)
4. Systemic arterial hypertension with left ventricular hypertrophy
5. Pulmonary hypertension
6. Acute or chronic renal failure
7. Ascitic liver cirrhosis
8. Endocrine diseases (eg, primary hyperaldosteronism, Cushing's syndrome)
9. Paraneoplastic increases (eg, small-cell lung cancer)

The positive predictive value of increased NPs for cardiovascular diseases, however, is somewhat limited. NPs are increased in all edematous disorders with a volume overload that lead to an increase in atrial tension or central blood volume (Table 3), including the last trimester of pregnancy. NPs are activated to their greatest extent in heart failure; however, they are also markedly increased in patients with renal failure, and in patients with ascitic liver cirrhosis and some endocrine disorders. Thus, increased NP concentrations do not always indicate cardiac diseases; however, an increase in NP is serious enough to warrant follow-up examinations in the respective patient.

Neurohormonal Activation in Heart Failure

The syndrome of heart failure is divided into congestive heart failure (CHF) and diastolic heart failure. Diastolic heart failure is characterized by impaired filling of one or both ventricles, with increased ventricular filling pressures and symptoms of heart failure despite normal systolic pump function. The key event leading to CHF is the loss of a critical amount of functioning myocardium after damage to the heart (for example, by AMI, toxins, infections, or prolonged cardiovascular stress). Although there is no completely adequate definition, from a hemodynamic viewpoint CHF can be defined as a pathophysiological state in which an abnormal cardiac function is responsible for failure of the heart to pump blood at a sufficient rate to meet the metabolic needs of tissues, or else does so only from an elevated filling pressure. Therapeutic strategies based on an exclusively mechanical understanding of CHF aim at optimizing preload, decreasing afterload, and improving contractility, which improves symptoms. Interestingly, only in some instances does the correction of hemodynamic abnormalities improve survival, and some drugs that have beneficial effects on hemodynamics adversely affect long-term outcomes in CHF patients.[20] The recognition that hemodynamic abnormalities alone do not explain many of the features of heart failure led to the search for other pathophysiological processes, including neuroendocrine activation. Today heart failure is no longer considered a mere hemodynamic disorder because of injury to the heart.[21] Heart failure is a multisystem disorder that is due to a primary abnormality in cardiac function, resulting in a series of hemodynamic, renal, neural, and hormonal adaptations with the aim to improve the mechanical environment of the heart. Heart failure develops when the compensatory hemodynamic and neurohormonal mechanisms are overwhelmed or exhausted. Neurohormonal activation plays a critical role for progression of the disease and provides the basis for the development of novel therapeutic strategies that

Table 4

Neurohormonal Systems Activated in Chronic Heart Failure: Vasopressors and Counterbalancing Systems

Vasoconstrictor Hormones	Vasodilator Hormones
Epinephrine, norepinephrine, sympathetic nervous system	Natriuretic peptides
Renin-angiotensin-aldosterone system	Prostaglandins
Arginine vasopressin	Kallikrein-kinin system
Endothelin	Calcitonin gene-related peptide
	Adrenomedullin

have positive effects on disease progression, for example, patients with marked neurohormonal activation benefit from ACE inhibitor treatment.[22] Neurohormonal systems are activated by the initial injury to the heart, and long-term neurohormonal activation exacerbates the hemodynamic abnormalities or exerts a direct toxic effect on the myocardium.

Low-output heart failure leads to arterial underfilling that sets into motion a series of baroreceptor-mediated events to restore arterial circulatory integrity and thus to maintain the perfusion of vital organs.[23] The SNS and RAAS are activated, and arginine-vasopressin is released nonosmotically (Fig. 1 and Table 4). The clinical signs and symptoms of CHF largely result from neurohormonal activation leading to vasoconstriction and renal salt and water retention. The immediate effects are tachycardia, increased myocardial contractility, vasoconstriction with increased cardiac preload and afterload, and stimulation of thirst. Salt and water retention occur with a delay of several days. These baroreceptor-mediated reflexes are of clear benefit during the acute phase. Over the long term, however, these reflexes are deleterious, increase preload and afterload, and lead to cardiac remodeling, pulmonary edema, and hyponatremia. Thus, it is highly desirable that future drugs for heart failure exert both favorable hemodynamic and neurohormonal effects, because the abnormalities cannot be assessed simply by the measurement of pressure, volume, and flow. The interplay of these neurohormonal and hemodynamic forces defines the syndrome of heart failure. Neurohumoral changes occur with disease progression of heart failure.

Neurohormones in Asymptomatic LVD

NPs have an important role in maintaining the compensated state of heart failure and delaying the progression to overt heart failure.

Cardiac hypertrophy that usually accompanies heart failure leads to increased ventricular production of ANP and BNP, and their release is further stimulated by the stretching of failing atrial and ventricular myocardium. NP activation is a beneficial humoral response in asymptomatic LVD, and increased plasma concentrations of NP and their N-terminal prohormone fragments are frequently found in asymptomatic LVD. Therefore, these peptides have been proposed as potentially useful early markers of heart failure. Although the activity of the SNS is also increased in early heart failure, plasma levels of norepinephrine are within reference limits during this stage of the disease. Catecholamines are increased only in severe heart failure. Another limitation of the use of catecholamines for diagnostic purposes is their limited specificity. Similarly, the plasma concentrations of the hormones of the RAAS system are not increased in untreated patients with asymptomatic LVD because this system is completely suppressed by the NP. However, hormones of the RAAS may be increased secondarily by the treatment of patients with diuretics. Through a combination of ventricular dilatation and hypertrophy, and the activation of vasoconstrictor and vasodilator forces, a delicate hemodynamic and neurohormonal balance is achieved that restores normal cardiac function at minimum energy cost (Table 4 and Fig. 1). In asymptomatic LVD, the activation of the RAAS and renal SNS is inhibited by NPs, and NPs are important for maintaining renal perfusion and urine flow.

Neurohormones in Symptomatic LVD

Long-term activation of endogenous, positive inotropic, and stress-reducing mechanisms leads to a diminution of their favorable physiological effects. With the progression of the disease to overt heart failure, the NPs cannot sufficiently counteract and suppress the SNS and RAAS. As the disease process continues, the glomerular filtration rate decreases, and there is increased sodium reabsorption in the proximal tubule, with decreased sodium delivery to the collecting duct, the place of action of NP.[24] In patients with late heart failure, there is also a fall in renal blood flow. As a result, salt retention occurs, which further deteriorates cardiac function. Although ANP and BNP are synthesized by ventricles as well as by the atria in patients with overt heart failure, the magnitude of this response is inadequate. NP receptor downregulation does not seem to be of great importance in the development of NP resistance in the setting of heart failure, because therapeutically administered NPs in decompensated heart failure patients are still effective.[25] Finally, vasoconstrictor effects and salt and water retention predominate, and symptoms develop in this state of excess total body fluid volume in the presence of LVD. In overt heart failure, the hormones of the RAAS and norepinephrine are also increased in plasma. In addition to these circulating factors, there is an increased release of locally active vasoconstrictors produced by the vascular endothelium in CHF, for example, endothelin. The clinical diagnosis of overt heart failure is not a real problem for the clinician; however, neurohormonal markers are relevant for estimating prognosis and monitoring treatment success.

Analytical and Preanalytical Issues of NP Measurement

Assays

Competitive immunoassays and immunometric assays are available for the determination of NPs and their N-terminal prohormone fragments. Competitive immunoassays are generally used to measure small molecules such as peptides. With the application of competitive immunoassays, extraction and chromatographic purification of relatively large volumes of plasma are necessary to achieve the desired analytical sensitivity and precision of the assay. However, the extraction step reduces practicability and may at the same time decrease precision because of variable recovery. The analytical sensitivity, specificity, and precision of immunometric assays, however, are usually superior to competitive immunoassays. Immunometric assays work very well with unextracted plasma samples in the environment of a routine laboratory.[26] NT pro-ANP and NT pro-BNP are larger molecules compared to the C-terminal biologically active peptides, which makes it easier to use the immunometric assay design. This sandwich format uses two antibodies directed against two sterically remote epitopes of the analyte. Recently, other noncompetitive immunometric assays for the measurement of ANP or BNP have been developed as well. In addition, a whole-blood, rapid, quantitative, point-of-care test for BNP determination is in development and will soon be available for routine use. Antibodies used in ANP and BNP assays should be highly specific for the biologically active part of the peptides, that is, the amino acid sequence containing the disulfide bridge between the two cysteine residues (Fig. 2). Due to differences in clearances and their higher molecular mass, plasma concentrations of NT pro-ANP and NT pro-BNP are markedly higher than ANP and BNP concentrations. ANP has the shortest biological half-life (approximately 3 minutes), followed by BNP (18 minutes), and NT pro-ANP (approximately 1 hour). The exact biological half-life for NT pro-BNP has not been determined. By analogy with NT pro-ANP, NT pro-BNP has probably a longer biological half-life compared with BNP. Higher concentrations of circulating NT pro-ANP and NT pro-BNP (10 to 50 times) make it easier to develop immunoassays with high analytical precision and a low detection limit without the need for extraction of plasma samples or very long incubation times.[27]

It is important to stress that currently these assays are not standardized, that is, they are not yet calibrated against common standards. Depending on the assay format, the plasma extraction step may be requested. Consequently, the results obtained with assays from different manufacturers may differ markedly in measurements of the same samples for the same analyte. Therefore, decision limits derived from clinical studies are only valid for the assay used in that particular publication and must not be extrapolated to other assays. A particular problem lies in the comparison of study results on NT pro-ANP or NT pro-BNP, because antibodies are directed toward specific and different regions of the N-terminal prohormones. For example, assays have been used which recognize different sequences of NT pro-ANP, mainly NT pro-ANP 1–98, NT pro-ANP 1–30, or NT pro-ANP 31–67. Because NT pro-ANP 1–98 is the major circulating NT pro-ANP form, this may not necessarily influence the clinical assessment of the marker; it does, however, impair the comparison of results obtained with different assays. It is also important to note that cross-reactivities of assays with prohormones, related NP, and fragments of NP or NP prohormones may differ among assays, and that antibodies and assays used in published clinical studies are sometimes not very well characterized, which hampers the comparability of published data on the utility of NP.

In Vitro Stabilities

Reports on NP and N-terminal prohormone in vitro stabilities are sometimes conflicting. Differences in collecting tubes (plastic versus glass) account for some conflicting reports. When working with NP, glassware should be avoided or must be siliconized very carefully to avoid absorption of NP to the walls. Possibly due to adherence to platelets or inherent instability of the added synthetic peptide, the stabilities in samples spiked with exogenous NP are much lower than in samples with high endogenous NP, which is the clinically relevant condition. In addition, antibodies of different assays may differ in their affinities to detect the degradation products of NP. According to recently published studies, aprotinin has limited effects on the decrease in ANP concentrations in whole blood at room temperature, but endogenous ANP is stable at room temperature in ethylenediaminetetraacetic acid (EDTA) whole blood containing aprotinin for up to 3 hours.[28] The addition of aprotinin had some benefit for ANP stability where sample collection, processing to plasma, and freezing took place within 2 to 3 hours. However, aprotinin could not prevent the decline of ANP at room temperature over 2 to 3 days. A consistent finding of all published studies is the stability of NT pro-ANP in EDTA whole blood or EDTA plasma for 3 to 4 days at room temperature. BNP is stable in EDTA whole blood for 3 days at room temperature as well.[28] NT pro-BNP was fully stable in EDTA whole blood or EDTA plasma for 3 days.[27] These studies indicate that aprotinin adds little benefit to the stability of BNP, NT pro-BNP, and NT pro-ANP at room temperature and that the in vitro stabilities of these analytes are sufficient for routine use in clinical practice. Blood samples can be taken into standard EDTA-containing tubes and can be sent to the laboratory without special requirements for shipment. The stabilities of NT pro-ANP, BNP, and NT pro-BNP allow for the shipping of whole blood EDTA samples by mail with subsequent separation, freezing, and measurement of these peptides at an appropriate reference laboratory, and for the use of these novel markers in both hospitals and community practice.

Blood Sampling Conditions

Similar to blood sampling for the measurement of catecholamines or hormones of the RAAS, some guidelines must be followed for blood sampling for NP measurement. In previous research, blood sampling for NP measurement has been performed under standard conditions, usually in a supine position after 10 minutes of rest. The influences of body position and exercise must be considered. The position of the patient must always be the same in all follow-up investigations, either supine or sitting. Blood may be drawn only after a standardized period of rest. The strongest increases after exercise are found for ANP (Table 5).[29,30] Due to their longer biological half-life, BNP and particularly NT pro-ANP or NT pro-BNP are less affected by exercise and are less sensitive to rapid changes in hemodynamics (for example, caused by change in body posture) than are ANP. The significance of eating habits (dietary sodium intake) and possible diurnal fluctuations of NPs have not been fully investigated. However, according to current knowledge, BNP exhibits neither diurnal variation nor any day-to-day variation in patients with CAD or heart failure. Blood should be drawn, whenever possible, at approximately the same time of the day in all follow-up determinations, to be on the safe side. As is the case for many other analytes, the influences of medication on NP measurement are not fully

Table 5

Natriuretic Peptide Concentrations at Rest and Post Stress in Controls, in Patients with Asymptomatic Left Ventricular Dysfuncton, and in Patients with Symptomatic Left Ventricular Dysfunction

	ANP (ng/L)	β-ANP (ng/L)	NT pro-ANP (ng/L)	BNP (ng/L)
Controls				
Rest	54.3 (10.4–100.3)	71.5 (12.8–236.6)	903 (153–1733)	22.8 (1.3–68)
Post stress	164.7 (41.1–443.1)	167.1 (13.2–736.4)	1306 (467–2715)	30.9 (1.7–91.6)
NYHA I				
Rest	77.9 (12.2–341.1)	119.5 (3.8–723.2)	1071 (176–4248)	42.3 (3.8–190)
Post stress	215 (38.1–617.3)	285.8 (25.8–1500)	1869 (207–14844)	61.2 (10.1–227.3)
NYHA II				
Rest	128.9 (41.6–394.4)	141.4 (9–419)	1633 (511–4364)	72.7 (10.4–376.5)
Post stress	240.4 (39.5–586.4)	257.8 (29.1–1062.7)	2068 (653–5673)	92.1 (20.5–418.8)

Data are given as mean and as minimum and maximum (in parentheses). Data from our own laboratory are published in Reference 30. In each group, all investigated markers increased significantly during exercise. Concentrations of all investigated NP of NYHA II patients at rest and post stress differed significantly from respective values of controls. Only BNP concentration of NYHA I patients at rest and post stress differed significantly from values of controls. There were close correlations among ANP and NT pro-ANP at rest and post stress ($r = 0.75–0.85$), whereas the correlations between ANP and BNP were significant but less close ($r = 0.40–0.64$).
ANP = atrial natriuretic peptide; BNP = brain natriuretic peptide; NT pro-ANP = N-terminal pro-ANP. NYHA I indicates asymptomatic left ventricular dysfunction; NYHA II indicates symptomatic mild left ventricular dysfunction.

established so far.[31] Glucocorticoid treatment may increase NP. Loop diuretics reduce NP without altering LVEF. It is well established that the degree of neurohormonal activation in heart failure is lowered by ACE inhibitor treatment, and therefore blood samples should be drawn, ideally, before the start of heart failure treatment. However, almost all studies on the usefulness of NP for the detection of LVD were not corrected for diurnal fluctuations, differences in treatment, or renal dysfunction. This fact even strengthens the implications of these studies for routine clinical practice since apart from renal failure, these effects are not deemed to be of great significance. Although increased NPs have been observed during the last trimester of pregnancy and in the immediate puerperium, there are no significant variations of ANP and BNP concentrations throughout the menstrual cycle.[32,33] In summary, blood sampling for NP measurement should be performed under standard conditions, although these are sometimes hard to meet in daily clinical practice. It is necessary to wait for more detailed studies on pre-analytical influences on NP plasma concentrations, before less strict sampling conditions can be recommended. The circumstances under which blood sampling for measurements of NT pro-ANP or NT pro-BNP should be performed appear to be more favorable than that for their C-terminal counterparts.[29]

NP Measurement: Clinical Results

Reference Intervals

Due to the lack of assay standardization and differences in methodology (plasma extraction step), no generally applicable reference intervals for NP and their N-terminal prohormone fragments can be given at this time. The upper reference limit in apparently healthy individuals must be determined for each assay separately. A dependence on gender has not been noted, but the majority of investigators found a weak dependence of the upper reference limit on age. Mild renal and systolic cardiac dysfunction as well as cardiac hypertrophy and lower than normal diastolic cardiac function contribute to the elevation of NP concentrations in elderly patients who do not have overt heart failure.[34]

NPs in the Diagnosis of Systolic and Diastolic LVD

All NPs increase with the clinical severity of heart failure assessed according to the NYHA classification (Table 5). NPs also correlate with invasively measured hemodynamic parameters at rest and during exercise (Table 6).[35] Both BNP and NT pro-ANP have been reported to be significantly elevated in asymptomatic NYHA Class I patients.[30,36–39] In the vast majority of comparative studies on BNP and NT pro-ANP, BNP was superior for the diagnosis of asymptomatic and symptomatic LVD.[30,37–39] These recent studies all failed to reproduce the high diagnostic accuracy of NT pro-ANP observed in the study of Lerman et al.[36] In our own investigation[30] with a strict definition of NYHA Class I of all tested NPs, only BNP concentrations in patients with asymptomatic LVD were significantly higher than those of controls (Table 5). NYHA Class I patients had an exercise capacity comparable to that of controls; however, the physiological average increase in LVEF during

Table 6

Correlations of ANP, N-terminal pro-ANP, and BNP with Hemodynamic Parameters at Rest and at Maximal Workload on Ergometer

	ANP		NT pro-ANP		BNP	
	Rest	Exercise	Rest	Exercise	Rest	Exercise
RAP	0.69	0.64	0.75	0.59	n.s.	0.66
MPAP	0.72	0.78	0.58	0.63	0.54	0.52
PCWP	0.73	0.71	0.66	0.70	0.55	0.57
CI	n.s.	n.s.	n.s.	n.s.	−0.77	−0.84

Data from our own laboratory; 16 patients with symptomatic mild left ventricular dysfunction. ANP = atrial natriuretic peptide; BNP = brain natriuretic peptide; CI = cardiac index; MPAP = mean pulmonary artery pressure; n.s. = not statistically significant; NT pro-ANP = N-terminal pro-ANP; PCWP = pulmonary capillary wedge pressure; RAP = right atrial pressure.

exercise was missing. Apart from the LVEF, there were no significant differences from controls. In accordance with the majority of published studies, we found moderate, negative inverse correlations between NP and LVEF. The closest but still moderate correlation was found for BNP. Correlations were closer in patients with overt heart failure (BNP with LVEF: r = -0.61). This finding has recently been confirmed by Richards et al.,[40] and all tested NPs correlated closer with LVEF than did catecholamines.

BNP continues to emerge as the superior cardiac hormone for the detection of LVD. The overall diagnostic accuracy of BNP in these studies varied considerably, ranging from moderate to good. The spectrum of disease affects the diagnostic value of a given test. The studies that included unselected patients, symptomatic as well as asymptomatic, obtained better results (area under receiver operating characteristic [ROC] curve for BNP was 0.8 to 0.9) than those studies restricting inclusion to asymptomatic or mildly symptomatic patients with less severe degrees of LVD (ROC areas for BNP approximately 0.7). The key finding of our investigation[30] was that the diagnostic performance of BNP was moderate but significantly better than that of ANP forms. All ANP forms had a comparable and poor diagnostic performance for the detection of asymptomatic LVD. Postexercise measurement was not significantly better than measurement at rest, and the combination of various NPs was not significantly better than a single BNP measurement. In similar studies, the combination of NPs did not improve diagnostic accuracy either.[37–39] However, BNP is unlikely to be suitable for more general routine screening for completely asymptomatic LVD.[30] In an unselected population of more than 1000 outpatients, McDonagh et al.[38] demonstrated that BNP is superior to NT pro-ANP in the diagnosis of left ventricular systolic dysfunction, which was defined as an echocardiographic LVEF less than 30%. They concluded that BNP is useful for screening for heart failure in symptomatic or high-risk individuals and in helping primary care physicians to identify patients who require further investigation. Similarly, Yamamoto and coworkers[39] convincingly demonstrated in about 100 patients referred for cardiac catheterization that BNP is the best NP to detect systolic LVD (defined as LVEF $<45\%$) or left ventricular hypertrophy (>120 g/m^2) and to identify impaired diastolic left ventricular function, defined either as impaired relaxation in echocardiography (Doppler parameters of left ventricular filling) or an elevated left ventricular end-diastolic pressure. The areas under ROC curves for BNP in the detection of these abnormalities ranged from 0.7 to 0.9. BNP also performed very well in a health screening program for asymptomatic cardiac diseases.[41] The areas under ROC curves for the detection of 13 patients with some form of asymptomatic cardiac disease in 481 consecutive individuals visiting a check-up clinic for a multiphasic health screening test were 0.81 (ANP) and 0.94 (BNP), with a significant difference between both markers. In summary, BNP appears to be the marker of first choice for the detection of systolic or diastolic LVD.

First results on NT pro-BNP show that this marker is comparable to BNP for the detection of LVD.[42,43] BNP and NT pro-BNP values are highly correlated ($r=0.88$). The correlation of NT pro-BNP with LVEF was comparable to that observed for BNP. The magnitude of increase in heart failure compared with controls is higher than that of BNP, which suggests that NT pro-BNP may be a more discerning marker of early LVD than is BNP. Whether NT pro-BNP is superior to BNP cannot be definitively judged at the moment because sufficient comparative data are still lacking.

Comparative Prognostic Value of Neurohormones in Heart Failure

Clear prognostic indicators are essential for the management of heart failure. A prognostic laboratory marker that is easy to measure would be of great clinical importance. Traditional prognostic indicators are the severity of symptoms (eg, NYHA classification), exercise capacity, ejection fraction, the presence or absence of pulmonary hypertension, Doppler echocardiographic mitral valve inflow pattern, left ventricular size, volume shape and mass, arrhythmias, weight loss, serum sodium, and the newly investigated markers—serum uric acid and thyroid function.[44] It is well recognized that traditional treatments that focused on correcting symptoms of hemodynamic alteration may not be beneficial in the long term. Apart from alterations in hemodynamics, heart failure leads to the activation of endogenous neurohormonal systems and to disturbances in ventilatory and autonomic controls. Markers of these alterations (for example, neurohormones, autonomic dysfunction, breathing pattern, and ventilatory response to exercise) have gained interest as prognostic indicators in heart failure patients. Neuroendocrine activation occurs early in the course of heart failure and directly correlates with mortality.[45–47] High levels of neurohormonal factors such as NP, norepinephrine, renin, and endothelin 1 are associated with increased mortality rates in patients with advanced heart failure. Only plasma renin activity and ANP were independently predictive of mortality in the Survival And Ventricular Enlargement (SAVE) study.[46] In the neurohormonal subgroup analysis of Studies of Left Ventricular Dysfunction (SOLVD), only a high plasma norepinephrine concentration was a significant marker for prognosis in patients with asymptomatic LVD.[47] Tsutamoto and coworkers[48,49] demonstrated that of norepinephrine, angiotensin II, endothelin, ANP, and BNP, BNP was the best neurohormonal marker for assessing prognosis in patients with advanced heart failure as well as in patients with asymptomatic or minimally symptomatic LVD. BNP was the only marker that provided prognostic information on morbidity and mortality independent from hemodynamic markers (such as LVEF, left ventricular end-diastolic volume index and pressure, and pulmonary capillary wedge pressure). Patients with low plasma BNP had an excellent long-term prognosis. High BNP concentrations, on the contrary, are related to high mortality (approximately 60% in 3-year follow-up). Therefore, BNP is, together with other standard clinical variables, emerging as an important, simple, and cost-effective indicator in the outpatient setting, even in asymptomatic or minimally symptomatic LVD.

Effects of Treatment on Plasma Concentrations of NPs: Candidate Markers for Tailoring Therapy in Heart Failure

Excessive neurohormonal activation is a maladaptive response to myocardial damage that eventually contributes to the development of overt heart failure. In the CONSENSUS trial, plasma angiotensin II concentrations were significantly higher in nonsurvivors than in survivors with advanced heart failure, and the effect of enalapril was related to the reduction of plasma angiotensin II.[45] Several studies have demonstrated that it is possible to lower NP concentrations in heart failure patients by therapeutic interventions.[46,50] In the meantime, similar to other neurohormones, NPs are routinely measured to evaluate the effects of new drugs for heart failure treatment on neurohormonal activation compared with the standard therapy in clinical trials. NPs can be used to titrate therapy in heart failure patients. Titration of ACE therapy according to plasma BNP was associated with more powerful inhibition of the RAAS, a significant fall in heart rate, and increased exercise capacity when compared with empiric therapy.[50] Heart failure patients may benefit from ACE dosage increase that is tailored to BNP concentrations. Given the fact that high BNP concentrations are an indicator of a poor prognosis, this fall in BNP probably reflects improved prognosis, which, however, remains to be proven in large prospective clinical trials. NPs are also useful as prognostic indicators to help the clinician know when to intensify treatment in an individual patient.

NPs in Hypertension

NPs are moderately (by about 20% to 100%) increased in some but not all patients with systemic arterial hypertension.[39,51] Left ventricular mass increases and increased ventricular wall tension lead to increased expression of NP in the left ventricle. Ventricular relaxation abnormalities cause an increase in left ventricular end-diastolic pressure. Consequently, right atrial pressure and wall tension increase as well; this is a stimulus especially for ANP and NT pro-ANP release. In addition, decreased renal function contributes to NP activation in hypertension. Plasma BNP is considered a reliable indicator (better than ANP) of left ventricular hypertrophy even in asymptomatic patients.[39,51] BNP is increased in hypertensive patients with hypertrophy of left ventricle via left ventricular overload or depression of diastolic function. NP plasma concentrations are increased in patients with obstructive pulmonary disease and pulmonary hypertension, as well.[52]

NPs in Patients with CAD

It is almost impossible to separate the effects of CAD on circulating NP from the effects of LVD, since CAD is the major cause of heart failure. However, from experimental studies and clinical investigations in humans, it is well known that NPs are released from ischemic cardiomyocytes, and that hypoxia stimulates the release of NPs from myocardium. We found a significant net release of BNP during reperfusion of the human heart after cardioplegic cardiac arrest in uneventful coronary artery bypass grafting patients that was closely correlated with myocardial lactate production.[53] During this period, the heart-lung machine still completely took over all circulatory functions, and the heart was beating unloaded without contributing to the maintenance of systemic arterial pressure; therefore, changes in filling pressures were not the reasons for the observed BNP release. The release of NP is also enhanced during myocardial ischemia induced by percutaneous transluminal coronary angioplasty.[54] However, such an event is more likely related to LVD during ischemia rather than ischemia per se. NPs are greatly increased in AMI patients during the acute phase. Probably, this increase is due to several factors, including alterations in hemodynamics, ischemia, increased synthesis especially in the peri-infarct zone, or release from necrotic myocardium. The available data suggest that ischemia does not increase peripheral venous NP concentrations as long as ventricular or atrial dysfunction are not present.

NPs and Neurohormones as Prognostic Markers during the Subacute Phase of AMI

The activation of all major neurohormonal systems that control vascular tone and fluid balance occurs in the first 2 to 3 days after AMI.[46] The extent and time course of neurohormonal activation in AMI depend on the degree of left ventricular impairment. Transient neurohormonal activation within the first 2 to 3 days is typical for the uncomplicated AMI, but persistent activation has been associated with manifest heart failure, morbidity, and mortality.[55] Low-grade neurohormonal activation including the RAAS may persist for 1 to 2 weeks after the infarct unless diuretic agents are given for treatment of heart failure. However, plasma norepinephrine and ANP remain elevated in proportion to the severity of LVD even in the absence of treatment.[46] The highest ANP values are found early after the onset of infarction, usually on admission, followed by a subsequent decline and an increase.[56] ANP concentrations on day 2 or 3 are related to LVEF. By contrast, BNP values peak approximately 16 hours after admission and decrease thereafter. In patients with LVD, a second peak is found during the subacute phase several days after AMI.[55,57] Interestingly, it has been demonstrated that the majority of patients in whom ACE inhibitor treatment is started early belong to the group of patients with a single early peak.[58] Of NP plasma, BNP most closely correlates with LVEF in AMI patients, and more closely than the other neurohormones as well. Plasma BNP measured within 1 to 4 days after AMI is a powerful, independent predictor of left ventricular function, heart failure, or

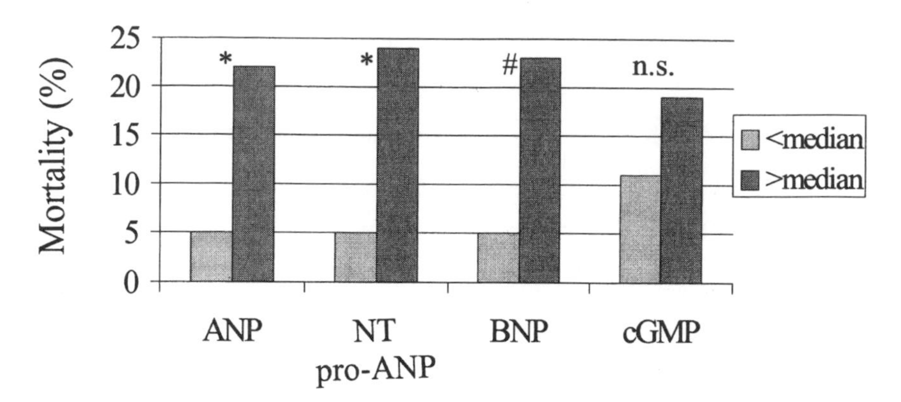

Figure 3. Neuroendocrine prediction of mortality after acute myocardial infarction. The 600-day mortality during follow-up of acute myocardial infarction patients is shown after grouping by median values during the subacute phase of infarction. ANP = atrial natriuretic peptide; BNP = brain natriuretic peptide; cGMP = cyclic guanosine monophosphate; n.s. = nonsignificant; NT pro-ANP = N-terminal pro-ANP. $^{*}P<0.05$; $^{\#}P<0.001$. Data from Reference 59.

death during the subsequent 14 months, and it is superior to ANP, NT pro-ANP, cGMP, and plasma catecholamines (Fig. 3).[59] In a comparison of plasma catecholamines, ANP, NT pro-ANP, BNP, and NT pro-BNP, the best predictors of prognosis in AMI patients were BNP and NT pro-BNP.[55] BNP and NT-BNP measured 2 to 4 days after AMI were comparably useful and independent and were additive prognostic markers to LVEF in the prediction of 2-year survival.

Therapeutic Potential of the NP System in Cardiovascular Diseases

Resistance to NP effects occurs in chronic heart failure. Nonetheless, short-term infusions of ANP and BNP in pharmacologic doses in heart failure patients have beneficial effects that lead to improved central hemodynamics with an increase in stroke volume.[25] Infusions of ANP and BNP also decrease plasma renin and aldosterone concentrations and increase sodium and water excretion. NP infusions reduce blood pressure and promote sodium excretion in patients with essential hypertension, as well.[60] There has been little success, however, with orally active analogues of NP. Studies with orally active NEP inhibitors (eg, candoxatrilat) have shown beneficial results in heart failure patients.[61] However monotherapy with these agents is not considered sufficiently potent to compete commercially with conventional diuretics,[62] although, in contrast to diuretics, NEP inhibitors do not activate the RAAS. Clinical studies using these agents have reported limited effects on blood pressure in hypertensive patients. Because available NEP inhibitors also impair the degradation of angiotensin II when administered in higher doses, recent research has focused on dual inhibitors of both the ACE and the NEP or the NEP and the endothelin-converting enzyme.[62,63] Manipulation of the NP environment, if successful, may form the basis for new strategies for the treatment of cardiovascular diseases.

Conclusions

ANP, NT pro-ANP, BNP, and NT pro-BNP do not provide identical, but rather complementary or even different, physiological information. Intracardiac pressures and atrial and ventricular wall tension are the prime regulators of cardiac NP release. ANP primarily reflects atrial overload, and BNP primarily reflects ventricular overload. From a pathophysiological point of view, the choice of the NP to be assayed depends on the pathophysiological goal of NP measurement in the individual patient. However, based on superior practicability for routine measurement and available comparative clinical investigations of all NPs, BNP emerges as the superior diagnostic and prognostic marker in patients with suspected systolic or diastolic LVD or heart failure as well as in patients during the subacute phase of AMI, or for screening hypertensive patients for the presence of left ventricular hypertrophy. BNP is clearly superior to other neurohormones for these purposes as well. The recently introduced marker NT pro-BNP is a promising new analyte that may eventually challenge BNP.

For primary care physicians, BNP measurement is useful, even in minimally symptomatic patients, to decide which patient with suspected LVD warrants further cardiological investigation, particularly when assessment of LVEF by echocardiography or radionuclide ventriculography is not readily available. BNP has an excellent negative predictive value, particularly in high-risk patients. In patients with normal BNP, a noncardiac disease should be considered. Patients with increased NP benefit diagnostically and therapeutically from echocardiographic evaluation of cardiac structure and function. It must be clearly stated that BNP cannot replace imaging techniques in the diagnosis of heart failure because these methods provide different information, and all patients with heart failure must be sent for echocardiography to further clarify the underlying cause of heart failure. The available data demonstrate the usefulness of BNP for selecting patients for further cardiological evaluation. The possible diagnostic role of NPs

Table 7

The Diagnostic Role of Natriuretic Peptides in the Context of Other Diagnostic Methods for Heart Failure

	Supports	Opposes
Appropriate signs	+ + +	+ (if absent)
Response to therapy	+ + +	+ + + (if absent)
ECG		+ + + (if normal)
Cardiac dysfunction on imaging	+ + +	+ + + (if normal)
Chest x-ray	if pulmonary congestion or cardiomegaly	+ (if normal)
Natriuretic peptides	+ + (if increased)	+ + + (if normal)

Grading of importance for diagnosis: + = some; + + + = great. The presence of appropriate symptoms is necessary for heart failure diagnosis.

in the context of other available diagnostic methods is shown in Table 7. The presence of symptoms is necessary for heart failure diagnosis. Appropriate clinical signs and response to treatment strongly support the heart failure diagnosis. Radiological signs of heart failure are not early markers. A patient with a normal resting electrocardiogram has a very low likelihood of having LVD. Imaging techniques (eg, echocardiography) are a cornerstone of heart failure diagnosis. BNP, if normal, has a very high negative predictive value, and a patient with a normal resting electrocardiogram and normal BNP has an extremely low likelihood of having ventricular dysfunction. However, the positive predictive value of increased BNP is somewhat limited. NPs are increased in all edematous disorders with a volume overload that lead to an increase in atrial tension or central blood volume (Table 3), but NPs are activated to their greatest extent in heart failure. Thus, increased NP concentrations do not always indicate heart failure; however, a confirmed increase in BNP is serious enough to warrant follow-up examinations in the patient. BNP, however, does provide information on neurohormonal activation in heart failure that is independent of and additive to the prognostic value of hemodynamic variables. NPs reflect both systolic and diastolic dysfunction, and both are important for prognosis. Therefore, BNP is also helpful to the cardiologist for monitoring the adequacy of current therapy and the disease course in heart failure patients and for estimating prognosis in heart failure and AMI patients.[64] For example, BNP could be helpful for selecting candidates for ACE therapy after AMI.[65]

There is now sufficient evidence to encourage physicians to gain experience with NPs as a supplement in the diagnostic evaluation of patients suspected of having heart failure, and in the assessment of prognosis in patients with heart failure. For high-risk patients, the measurement of BNP is at least as cost effective for the screening of heart failure as is the use of prostate-specific antigen, mammography, or cervical smears for the screening of carcinoma.[66] The assessment of BNP is simple and can be repeated easily during follow-up, and thus is a useful addition to the standard clinical investigation of patients with LVD. In addition, the therapeutic potential of NPs remains attractive, and drugs that augment their biological activity may yet enter our therapeutic armamentarium in heart failure patients.

References

1. Crowie MR, Moster DA, Wood DA, et al. The epidemiology of heart failure. *Eur Heart J* 1997;18:206–215.
2. Sharpe N, Doughty R. Epidemiology of heart failure and left ventricular dysfunction. *Lancet* 1998;352(suppl I):3–7.
3. Crowie MR, Wood DA, Coats AJS, et al. Incidence and aetiology of heart failure. *Eur Heart J* 1999;20:421–428.
4. Mosterd A, Hoes AW, de Bruyne MC, et al. Prevalence of heart failure and left ventricular dysfunction in the general population—The Rotterdam study. *Eur Heart J* 1999;20:447–455.
5. Mc Donagh TA, Morrison CE, Lawrence A, et al. Symptomatic and asymptomatic left-ventricular dysfunction in an urban population. *Lancet* 1997; 350:829–833.
6. Packer M, Cohn JN. Consensus Recommendations for the management of chronic heart failure. *Am J Cardiol* 1999;83:1A-32A.
7. Stevenson LW, Perloff JK. The limited reliability of physical signs for estimating hemodynamics in chronic heart failure. *JAMA* 1989;261:884–888.
8. McKee PA, Castelli WP, McNamara PM, et al. The natural history of congestive heart failure: The Framingham study. *N Engl J Med* 1971;285:1441–1448.
9. de Bold AJ, Borenstein HB, Veress AT, et al. A rapid and potent natriuretic response to intravenous injection of atrial myocardial extracts in rats. *Life Sci* 1981;28:89–94.
10. Koller KJ, Goeddel DV. Molecular biology of the natriuretic peptides and their receptors. *Circulation* 1992;86:1081–1088.
11. Levin ER, Gardner DG, Samson WK. Natriuretic Peptides (review). *N Engl J Med* 1998;339:321–328.
12. Wilkins MR, Redondo J, Brown LA. The natriuretic-peptide family. *Lancet* 1997;349:1307–1310.
13. Burrell LM, Lambert HJ, Baylis BH. Effect of atrial natriuretic peptide on thirst and arginine vasopressin release in humans. *Am J Physiol* 1991;260: R475-R479.
14. Hunt PJ, Espiner EA, Nicholls MG, et al. Differing biological effects of equimolar atrial and brain natriuretic peptide infusions in normal man. *J Clin Endocrinol Metab* 1996;81:3871–3876.
15. Sudoh T, Kangawa K, Minamino N, et al. A new natriuretic peptide in porcine brain. *Nature* 1988;332:78–81.

16. Stingo AJ, Clavell AL, Aarhus LL, Burnett JC. Cardiovascular and renal actions of C-type natriuretic peptide. *Am J Physiol* 1992;262:H308-H312.
17. Matsukawa N, Grzesik W, Takahashi N, et al. The natriuretic peptide clearance receptor locally modulates the physiological effects of the natriuretic peptide system. *Proc Natl Acad Sci U S A* 1999;96:7403–7408.
18. Saito Y, Nakao K, Arai H, et al. Augmented expression of atrial natriuretic polypeptide gene in ventricle of human failing heart. *J Clin Invest* 1989;83: 298–305.
19. Wei CM, Kao PC, Lin JT, et al. Circulating β-atrial natriuretic factor in congestive heart failure in humans. *Circulation* 1993;88:1016–1020.
20. Brozena S, Jessup M. Pathophysiologic strategies in the management of congestive heart failure. *Annu Rev Med* 1990;41:65–74.
21. Packer M. The neurohumoral hypothesis: A theory to explain the mechanism of disease progression in heart failure. *J Am Coll Cardiol* 1992;20: 248–254.
22. Francis GS, Cohn JN, Johnson G, et al. Plasma norepinephrine, plasma renin activity and congestive heart failure: Relations to survival and the effects of therapy in V-HeFT-II. *Circulation* 1993;87(suppl VI):VI40-VI48.
23. Schrier RW, Abraham WT. Hormones and hemodynamics in heart failure. *N Engl J Med* 1999;341:577–585.
24. Harris PJ, Thomas D, Morgan TO. Atrial natriuretic peptide inhibits angiotensin-stimulated proximal tubular sodium and water reabsorption. *Nature* 1987;326:697–698.
25. Yoshimura M, Yasue H, Morita E, et al. Hemodynamic, renal, and hormonal response to brain natriuretic peptide infusions in patients with congestive heart failure. *Circulation* 1991;84:1581–1588.
26. Puschendorf B, Mair J. Cardiac diseases. In Thomas L (ed): *Clinical Laboratory Diagnostics—Use and Assessment of Clinical Laboratory Results*. Frankfurt, Germany: TH Books; 1998:101–119.
27. Karl J, Borgya KA, Gallusser A, et al. Development of a novel, N-terminal-proBNP (NT-proBNP) assay with a low detection limit. *Scand J Clin Lab Invest* 1999;59(suppl 230):177–181.
28. Buckley MG, Marcus NJ, Yacoub M. Cardiac peptide stability, aprotinin and room temperature: Importance for assessing cardiac function in clinical practice. *Clin Sci* 1999;97:689–695.
29. Wijbenga JAM, Balk AHMM, Boomsma F, et al. Cardiac peptides differ in response to exercise: Implications for patients with heart failure in clinical practice. *Eur Heart J* 1999;29:1424–1428.
30. Friedl W, Mair J, Thomas S, et al. Natriuretic peptides and cyclic guanosine 3′,5′-monophosphate in asymptomatic and symptomatic left ventricular dysfunction. *Heart* 1996;76:129–136.
31. Clerico A, Iervasi G, Mariani G. Pathophysiological relevance of measuring the plasma levels of cardiac natriuretic peptide hormones in humans. *Horm Metab Res* 1999;31:487–498.
32. Rutherford AJ, Anderson JV, Elder MG, et al. Release of atrial natriuretic peptide during pregnancy and immediate puerperium. *Lancet* 1987;I: 928–929.
33. Maffei S, Clerico A, Vitek F, et al. Circulating levels of cardiac natriuretic peptides during menstrual cycle. *J Endocrinol Invest* 1999;22:1–5.
34. Sayama H, Nakamura Y, Saito N, et al. Why is the concentration of plasma brain natriuretic peptide in elderly inpatients greater than normal? *Coron Artery Dis* 1999;10:537–540.

35. Friedl W, Mair J, Thomas S, et al. Relationship between natriuretic peptides and hemodynamics in patients with heart failure at rest and after ergometric exercise. *Clin Chim Acta* 1999;281:121–126.
36. Lerman A, Gibbons RJ, Rodeheffer RJ, et al. Circulating N-terminal atrial natriuretic peptide as a marker for symptomless left-ventricular dysfunction. *Lancet* 1993;341:1105–1109.
37. Cowie MR, Struthers AD, Wood DA, et al. Value of natriuretic peptides in assessment of patients with possible new heart failure in primary care. *Lancet* 1997;350:1347–1351.
38. McDonagh TA, Robb SD, Murdoch DR, et al. Biochemical detection of left-ventricular dysfunction. *Lancet* 1998;351:9–13.
39. Yamamoto K, Burnett JC, Jougasaki M, et al. Superiority of brain natriuretic peptide as a hormonal marker of ventricular systolic and diastolic dysfunction and ventricular hypertrophy. *Hypertension* 1996;28:988–994.
40. Richards AM, Nicholls MG, Yandle TG, et al. Neuroendocrine prediction of left ventricular function and heart failure after acute myocardial infarction. *Heart* 1999;81:114–120.
41. Niinuma H, Nakamura M, Hiramori K. Plasma B-type natriuretic peptide measurement in a multiphasic health screening program. *Cardiology* 1998: 90:89–94.
42. Hunt PJ, Richards AM, Nicholls MG, et al. Immunoreactive amino-terminal pro-brain natriuretic peptide (NT-proBNP): A new marker of cardiac impairment. *Clin Endocrinol* 1997;47:287–296.
43. Talwar S, Squire B, Davies JE, et al. Plasma N-terminal pro-brain natriuretic peptide and the ECG in the assessment of left-ventricular systolic dysfunction in a high risk population. *Eur Heart J* 1999;20:1736–1744.
44. Piepoli M. Diagnostic and prognostic indicators in chronic heart failure [editorial]. *Eur Heart J* 1999;20:1367–1369.
45. Swedberg K, Eneroth P, Kjekshus J, et al. Hormones regulating cardiovascular function in patients with severe congestive heart failure and their relation to mortality. *Circulation* 1990;82:1730–1736.
46. Rouleau JL, Packer M, Moye L, et al. Prognostic value of neurohumoral activation in patients with an acute myocardial infarction: Effect of captopril. *J Am Coll Cardiol* 1994;24:583–591.
47. Benedict CR, Shelton B, Johnstone DE, et al, for the SOLVD Investigators. Prognostic significance of plasma norepinephrine in patients with asymptomatic left ventricular dysfunction. *Circulation* 1996;94:690–697.
48. Tsutamoto T, Wada A, Maeda K, et al. Attenuation of compensation of endogenous cardiac natriuretic peptide system in chronic heart failure: Prognostic role of plasma brain natriuretic peptide concentration in patients with chronic symptomatic left ventricular dysfunction. *Circulation* 1997;96: 509–516.
49. Tsutamoto T, Wada A, Maeda K, et al. Plasma brain natriuretic peptide level as a biochemical marker of morbidity and mortality in patients with asymptomatic or minimally symptomatic left ventricular dysfunction—comparison with angiotensin II and endothelin-1. *Eur Heart J* 1999; 20:1799–1807.
50. Murdich DR, McDonagh TA, Byrne J, et al. Titration of vasodilator therapy in chronic heart failure according to plasma brain natriuretic peptide concentration: Randomized comparison of the hemodynamic and neuroendocrine effects of tailored versus empirical therapy. *Am Heart J* 1999;138: 1126–1132.

51. Yasumoto K, Takata M, Ueno H, et al. Relation of plasma brain and atrial natriuretic peptides to left ventricular geometric patterns in essential hypertension. *Am J Hypertens* 1999;12:921–924.
52. Lang CC, Coutie WJ, Struthers AD, et al. Elevated levels of brain natriuretic peptide in acute hypoxaemic chronic obstructive pulmonary artery disease. *Clin Sci* 1992;83:529–533.
53. Mair P, Mair J, Bleier J, et al. Augmented release of brain natriuretic peptide during reperfusion of the human heart after cardioplegic cardiac arrest. *Clin Chim Acta* 1997;261:57–68.
54. Ikaheimo MJ, Ruskoaho HJ, Airaskinen KEJ, et al. Plasma levels of atrial natriuretic peptide during myocardial ischemia induced by percutaneous transluminal coronary angioplasty or dynamic exercise. *Am Heart J* 1989; 117:837–841.
55. Richards AM, Nicholls MG, Yandle TG, et al. Plasma N-terminal pro-brain natriuretic peptide and adrenomedullin: New neurohormonal predictors of left ventricular function and prognosis after myocardial infarction. *Circulation* 1998;97:1921–1929.
56. Wencker M, Lechleitner P, Dienstl F, et al. Early decrease of atrial natriuretic peptide in acute myocardial infarction. *Lancet* 1987;1:1369.
57. Morita E, Yasue H, Yoshimura M, et al. Increased plasma levels of brain natriuretic peptide in patients with acute myocardial infarction. *Circulation* 1993;88:82–91.
58. Mizuno Y, Yasue H, Oshima S, et al. Effects of angiotensin-converting enzyme inhibitor on plasma B-type natriuretic peptide levels in patients with acute myocardial infarction. *J Card Fail* 1997;3:287–293.
59. Richards AM, Nicholls MG, Yandle TG, et al. Neuroendocrine prediction of left ventricular function and heart failure after acute myocardial infarction. *Heart* 1999;81:114–120.
60. Janssen WMT, de Zeeuw D, van der Hem GK, et al. Antihypertensive effect of 5-day infusion of atrial natriuretic factor in humans. *Hypertension* 1989; 13:640–646.
61. Munzel T, Kurz S, Holtz J, et al. Neurohormonal inhibition and hemodynamic unloading during prolonged inhibition of ANF degradation in patients with severe chronic heart failure. *Circulation* 1992;86:1089–1098.
62. Evans RR, DiPette DJ. New or developing antihypertensive agents. *Curr Opin Cardiol* 1997;12:382–388.
63. Fournie-Zaluski MC, Gonzalez W, Turcaud S, et al. Dual inhibition of angiotensin-converting enzyme and neutral endopeptidase by the orally active inhibitor mixanpril: A potential therapeutic approach in hypertension. *Proc Natl Acad Sci U S A* 1994;91:4072–4076.
64. Struthers AD. How to use natriuretic peptide levels for diagnosis and prognosis. *Eur Heart J* 1999;20:1374–1375.
65. Motwani JG, McAlpine H, Kennedy N, et al. Plasma brain natriuretic peptide as an indicator of ACE inhibition after myocardial infarction. *Lancet* 1993;341:1109–1113.
66. Dickstein K. Natriuretic peptides in detection of heart failure [editorial]. *Lancet* 1998;351:4.

Index